BUILD A
DRONE

BUILD A DRONE

A STEP-BY-STEP GUIDE TO DESIGNING, CONSTRUCTING, AND FLYING YOUR VERY OWN DRONE

BARRY DAVIES, BEM

Skyhorse Publishing

Skyhorse Publishing books may be purchased in bulk at special discounts for sales promotion, corporate gifts, fund-raising, or educational purposes. Special editions can also be created to specifications. For details, contact the Special Sales Department, Skyhorse Publishing, 307 West 36th Street, 11th Floor, New York, NY 10018or info@skyhorsepublishing.com.

Skyhorse® and Skyhorse Publishing® are registered trademarks of Skyhorse Publishing, Inc.®, a Delaware corporation.

Visit our website at www.skyhorsepublishing.com.

10 9 8 7 6

Library of Congress Cataloging-in-Publication Data is available on file.

Cover design by Tom Lau
Cover photos courtesy of the author

Print ISBN: 978-1-51070-705-4
Ebook ISBN: 978-1-51070-706-1

Printed in China

To my daughter, Grace, who turned out to be a great drone flyer.

DISCLAIMER

The information in this book is for educational and entertainment purposes only. Although the author and publisher have made every effort to ensure that the information in this book was correct at press time, the author and publisher do not assume and hereby disclaim any liability to any party for any loss, damage, or disruption caused by errors or omissions, whether such errors or omissions result from negligence, accident, or any other cause.

TABLE OF CONTENTS

EDITOR'S NOTE

During the editorial process of this book, author Barry Davies passed away. He was seventy-one years old.

As someone who had the pleasure of working with Mr. Davies on several projects, I can say that his knowledge was only equaled by his kind heart and sense of humor.

I am honored to have been able to collaborate with such an incredible person who taught me more than I could have ever imagined.

We will miss you, Barry.

INTRODUCTION

It's a small plastic toy that we would once call a model airplane, but add the military veneer of "drone" and the toy becomes a lot more evolved than just a model airplane. So when toys become sophisticated enough to be called drones, the name *drone* creates something that is currently conceived as a liability.

I have been flying for more years than I can remember, a skill that was rekindled when I brought my first Microdrone from Germany. Its simplicity of design and ease of control encouraged me to start a new division within the company BCB Robotics. For the past nine years I have been immersed in every aspect of aerial robotics, while watching my personal collection of purchased and hand-made drones grow dramatically. Like so many others, I have fallen in love with drones. *Build a Drone* means just that; I built, calibrated, tuned, and flew the drone build in this book in less than three hours.

For many people, the term *drone* seems to conjure images of military use and war weaponry. As a result, the mere thought of these futuristic flying devices tends to pose security and privacy concerns. While nothing could be further from the truth, years ago you could hardly hear the word *drone* outside of military circles; now it is spoken worldwide. Not because of the military but because of the explosion in the sale of model drones.

Drone sales are not just hot, they are on fire. Forget the military use of drones for a moment; civilian drone sales are set to top $40 billion worldwide, creating more than 50,000 jobs in the US alone. The proliferation of unmanned aircraft is already outpacing the regulations that govern them; both the US FAA and the UK's CAA are falling behind the rapid development in drone technology that is driven by hundreds of thousands of young innovators worldwide.

This is despite the fact that the media seem to jump on the bandwagon every time some idiot with a private drone flies close to an airfield and puts people's lives in danger, and despite the cries of attacks on individual privacy. Yet their advantages are numerous. A drone is an eye in the sky that can detect and help stop crime. Drones were used during the recent earthquake in Nepal for damage assessment. They can fly over areas where humans dare not venture, such as minefields. They have rescued children from flooding and have been used for search and rescue operations in several countries. When used with the right intentions, drones are good.

If drones could let you know what they are good at (and that day is not far away) they would say they are designed for dull, dirty, and dangerous work. *Dull* may encompass looking at crop growth and making the farming industry more efficient so we can feed the planet. *Dirty* could mean going into places inaccessible to humans, and *dangerous* work might be assessing the radiation from a leaking nuclear plant. Drones can and will be extremely useful to society as a whole. For the moment, we only hear of military drones bombing some far-off target, or how some kid flew his drone over an airport, or about the idiot who used a shotgun to shoot down a drone. This is today's news; tomorrow it will be different when drones start saving lives.

One thing is for sure: drones are here to stay.

So you are thinking about owning a drone? Well, you have a choice: buy one ready made, or purchase one in kit form—or you can follow the instructions in this book and make your own. Whichever path you choose, remember that with a drone comes responsibility; it may only be a "toy," but it's a toy that flies in the air. No matter what size or type of drone you possess, you have to take into consideration some basic truths. Drones can become an obsession and take over much of your free time. And owning or building a drone does not grant you an automatic pilot's license—it is equally as important to learn to fly safely.

The main aim of this book is to give you a step-by-step guide on how to build a drone without all the technical jargon (although you will learn some technical terms as you progress). Additionally, the book will show you how the basic drone can be upgraded so that you can fly autonomously, using way-points, or set the drone to Follow Me mode. You will also be able to improve the look of your drone, making it more professional.

Build a Drone contains a vast amount of information about all forms of drones, from those used by the military and commercial markets to the models available for hobbyists. It also explains their use within society and how they will enhance mankind. Most of all, it highlights the vital aspects of safety required when flying a drone, mainly the FAA and CAA regulations that discipline its use.

If you can read, then you can build a drone. You don't need a degree in aeronautical engineering. True, there is some programming involved, and fine adjustments will make your drone fly better, but they are easy to understand.

How much will it cost? To be perfectly honest, it is cheaper to buy a drone than to build one. Most new drones come out of the box almost ready to fly after some basic assembly and charging the battery. However, if you want to understand the drone, you are better off building your own—added to which is great fun. The basic drone shown in this book will cost you around $300, but if you want to remodel, adding more sophistication such as telemetry and LIDAR, you can count on doubling that.

Build a Drone contains insight into all the types of drones available and what they are used for. It is also a guide to the advantages of buying versus building. For those that choose the latter, the book explains the mechanical parts required to keep a drone stable in the air and how they control its actions and functions in flight.

Build a Drone is more than a simple step-by-step guide. It explains how to put the parts together, calibrate, and adjust settings to get your bird into the air safely. The book covers every aspect of obtaining your own drone, from simple construction to the workings of a Ground Control System, and using software to control it. It also shows how to take your basic drone and turn it into a thing of beauty by designing your own airframe.

To build a drone, you only need to know how to read; the rest will come naturally. If you learn what a component does and grasp its function within the drone's structure, slowly but surely you will understand what makes a drone fly stable in the air. You need only to gather your component parts and assemble them, adding the firmware and software to make it fly and carry out your commands. None of this requires programming skills or advanced knowledge of electronics.

Why would you want a personal drone? Well, there are many possible reasons for having your own drone. Most of us love things that fly, and up until a few years ago this would mean going to an aero-model club and learning to fly a fixed-wing aircraft or a helicopter. Now we have drones and things are much simpler, especially when it comes to flying. Additionally, you may want to use your drone for a more commercial application like aerial mapping or photography and become a professional flyer earning your living from flying drones. No matter the reason for wanting a drone, *Build a Drone* will help you decide what is best for you personally.

To build or buy your own drone means investing in some serious equipment and personal time, both of which can be expensive. When you finally have your drone, learning to fly it is the key element to minimizing costly crashes and damage. Although quad-copters can literally fly themselves when in automatic mode, if they're not calibrated correctly and you have not installed a range of fail-safe protocols, you could still lose your drone. Worse still, it could cause damage to people or property. Learning to fly safely is paramount to protecting your drone investment.

Another worry is where to fly your drone; you may well own one but is it legal to fly? Even when it is legal, there are considerations such as privacy and trespass. Aviation law on flying a drone is different around the world and even from state to state in America, so you are best advised to consult the law in your area. Visit a local flying club and ask the advice of those who have been safely flying for years.

As you read through *Build a Drone*, the answers to all of the above questions and advice will become clear, especially regarding construction. The one thing you might find a little confusing is all the acronyms throughout the book, such as LIDAR (Light Detection and Ranging). Don't worry too much, as most are explained throughout the book and those that are not can be found in the Glossary. Additionally, I would advise all those truly interested in any form of aerial robotics to search through the brilliant ArduPilot website: ardupilot.com/

In a world that is rapidly changing, I personally find that designing and building a drone is very relaxing and extremely satisfying. So enjoy your drone build and get it flying safely. It is a new and exciting hobby, one that reaches out to advanced technology in a way that all of us can understand.

You may have been building model aircraft and flying at your local club for years; you may be an engineer with an interest in drone technology; or you may have purchased a drone for a hobby. No matter what role you play, you are a partner in the drone revolution. The day when drones are common in the skies overhead, you can say to your grandchildren, "I built one of those."

Chapter 1

WHAT IS A DRONE?

watched the news this morning as they announced the death of Jihadi John, believed killed in an American drone strike. Real name Mohammed Emwazi, he become prominent after appearing in a series of gory propaganda videos showing the beheading of several British, American, and Japanese captives. Jihadi John was targeted in a vehicle as he left the ISIS stronghold of Raqqa in eastern Syria. Imagine a flying machine, which could have flown from the United States, able to locate and kill a man thousands of miles away. Actually, I believe the drone may have been launched from the Ali Al Salem air base in Kuwait. The funny thing is, drone strikes are such a regular occurrence today that it's no longer considered news. Yet, by contrast, in the same month, a British filmmaker was fined $1,350 (£1,200) for flying his model drone over Hyde Park in London.

Less than a week later, 129 people were killed in a massive attack in Paris, France. The attacks were carried out by suicide gunmen who, after shooting down many of their victims, detonated their suicide vests killing even more. France was quick to respond, sending fighters to bomb some of the ISIS strongholds in Syria. Within a week, the UN Security Council unanimously approved an action that nations should take all necessary measures in the fight against ISIS. The US increased its drone strikes.

Surprisingly, drones have been around for some time, but the word *drone* seems to have suddenly been transformed into the Prince of Darkness . . . well maybe not quite that bad, but in the happy days of model making, building and flying a model plane was something done for pleasure. So let's step back a little in time before all the media hype about flying objects invading your privacy and how killer drones roam the skies looking for trouble. In reality, there is more of a risk from bird strikes on manned aircraft than any serious danger from drones, but bird strikes do not make headline news.

I have always called my personal drones Unmanned Aerial Vehicles (UAVs). However, in September 2015, I attended the conference at the main Naval Air Base in Cornwall, UK. Everyone was there, including the people who designed ScanEagle in America and the British police, who displayed a DJI Inspire. It was without doubt the best "across the board" drone conference I have ever attended. I was impressed with the open debate but silently amused to hear that the official word for any UAV was now *drone*. Military or civilian, large or small, in the UK it's a drone. Likewise, it's not *autonomous flying*, it's now called *automatic*, and so I will stick with these terms for the moment.

Author's Note: Throughout the book I may refer to "your drone" and "your quad-copters" in the same sentence. In both cases they are one and the same.

The use of drones can be classified into three different areas: military, civilian, and hobbyist. Although the use of drones in the military has been around for a long time, recent innovations in communication have rekindled the possibilities and benefit of drones upon the battlefield. As the face of war has changed, so has the weaponry designed to wage it. In the UK, for example the British military has drones at its disposal, the largest of these being the American-made Reaper. Britain also has the Watchkeeper, the Desert Hawk, and the small Black Hornet—the latter is just 18 grams (0.63 ounces).

On the civilian side, drones have a very good future, but not in the present form as most people are led to believe. The media shows parcel delivery by drone, and a hot meal being placed on a table by the flying restaurant service. Trust me, that is not going to happen—well, not for a long time yet. And why not? It's because of the "what if" factor. For example, our food delivery drone is working perfectly with thousands of flights a day to happy customers. Then one day some idiot on the table decides it would be fun to put a fork up into the propellers just to see what would happen; obviously the drone would lose balance and crash wildly causing serious damage to anyone in close proximity. Our intention was to deliver food, but we did not anticipate the "what if" factor—the idiot sitting at the table. So it will be a long time before drone delivery flights are considered totally safe.

The Black Hornet originated out of Norway and is the world's smallest military drone. At just 16 grams, it has a 1.6 kilometer digital data-link range with a 25-minute flight time. The whole system, including two drones and the controls, only weighs 1.3 kilos (45 ounces).

While the general public might refer to all drones in the same voice, the authorities see them quite differently, and as such classify them. In both the hobbyist and commercial worlds, *size* is the determining factor. A drone is classified in (mm) with the frame size being represented by the greatest point–to-point distance between two motors on a drone; yet you will often find that most propellers are sized in inches. However, when it comes to the government, both the UK and the US classify the type of drone by weight:

7 ounces (200 grams) or less is called Nano
7 ounces (200 grams) to 4½ pounds (2 kilograms) is called Micro
4½ pounds (2 kilograms) to 45 pounds (20 kilograms) is called Mini
45 pounds (20 kilograms) to 330 pounds (150 kilograms) is called Small

After this we move into the larger military drones, normally called Tactical, Male, and Hale, which is anything 330 pounds (150 kilograms) and above. More recently there has been a review of drones, and the military now classifies a

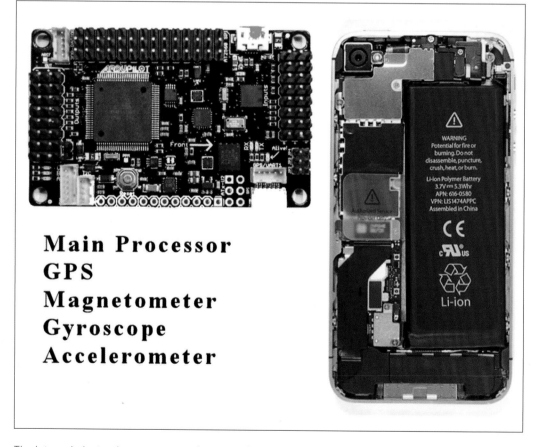

Main Processor
GPS
Magnetometer
Gyroscope
Accelerometer

The internal electronic components of a smartphone are similar to those found in the autopilot that is used for our drone build in this book.

drone according to its potential. For example, a 200 gram Nano quad-copter would not be so harmless if it had the ability to deliver an explosive device or become weaponized; then it would be classified right up there with the Predator.

So why have drones become so newsworthy? I personally think it's the sudden and rapid development of unmanned aviation. The driving force and contributing factors could be the technical innovations in aviation: the autopilot, the inertial navigation system, and data links to name a few. In the past, drone development was hindered by technological insufficiencies through most of the twentieth century; however, concentrated efforts in various military projects overcame the basic problems of automatic stabilization, remote control, and autonomous navigation by the sixties.

Microprocessors have become ever more capable in their ability, and are now smaller and more lightweight. Nano sensors and GPS were used to convert the mobile phone industry into a worldwide network, producing a handheld communications device that your could slip into your pocket. What has this got to do with drones? Well a smartphone and a drone share many similarities, minus the motors and propellers. Today's autopilots have a main processor, GPS, magnetometer, gyroscope, and accelerometer just like the phone in your pocket.

DRONE USAGE

The one thing about drones is that they use the open sky above and, for the most part, are out of reach and without hindrance; the major drawback is if they fail, they fall to earth. So, until we can develop drones that are considered totally safe, we must restrict them to areas where, if they fail, they will do no damage and cause no injury. Strangely enough, this still leaves us with lots of possibilities:

- Flying a drone over a wilderness searching for a missing person.
- Damage assessment of a devastated area after an earthquake.
- Search and rescue at sea.
- Helping farmers control the growth of crops.
- Checking landfill sites for leaking gasses.

None of those activities poses a problem should the drone crash and fall to earth, yet these areas account for most of the planet's surface. Once the media hype has died down, drones will become more widely acceptable. People may complain about the noise, but cars make more noise; it's a matter of acceptance. The old saying "what the eye does not see the heart does not grieve over" is very apt for drones. If our drone is flying high enough not to be heard, then almost certainly no one will bother; it's only when the drone is low enough for people to hear that they will look up.

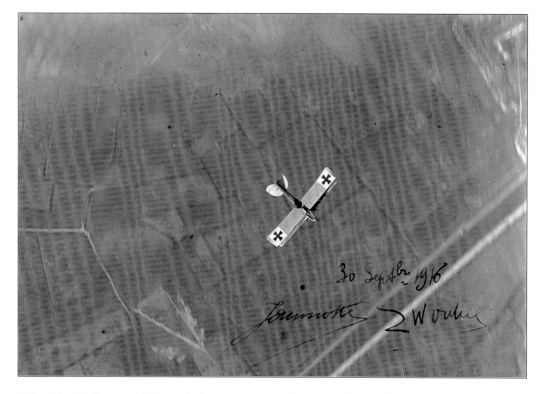

This photo of a German Albatross C.III reconnaissance plane was taken in 1916 from a Belgian plane. This was both strategically and technically difficult, in addition to the inherent danger of photographing.

Safety and media hype behind us for the moment, let us look at how we arrived at present-day drones. On August 22, 1849, Austria used balloons to drop explosives on Venice. While the city was under siege, one bright artillery officer had the idea of sending balloons with a bomb attached that would explode when over Venice. As the prevailing winds were from the sea, the balloons were actually launched from a ship called the *Volcano*. Many of the balloons did explode over Venice with devastating effect, not just in collateral damage but by terrifying the people, as the bombs were filled with shrapnel.

In the American Civil War, both sides used hot-air balloons to monitor each other's movements and occasionally to direct troops or adjust artillery fire. In 1898, during the Spanish-American War, the US used a kite with a camera attached to it, thus creating the first aerial surveillance. Both British and German planes of the First World War used cameras to take excellent imagery of enemy positions.

However, the first true drones appeared around 1941, when German scientists experimented with radio-controlled missiles during World War II, bombarding both British shipping at sea as well as strategic locations inland. They developed a method of dropping the missile from an aircraft, after which the pilot guided it to its target.

During 1941, the Naval Aircraft Factory developed an early unmanned combat aerial vehicle, called the TDN, referred to at the time as an "assault drone." It was tested in early 1942, and by March the government had ordered ten units. Expensive to build, the TDN was quite sophisticated, with a television camera in the nose, and could be flown by remote control by pilots aboard chase aircrafts. It proved only partially successful, although the TDN did destroy some Japanese shipping in the Pacific. The final mission was flown on October 27, 1944, with a total of some 50 drones having been expended, of which 31 reached their target successfully.

They were, in fact, the drones of their day. In the fifties, America and Russia were competing to conquer space; the scientists on both sides had to figure out how to fly a rocket without a human on board, launching satellites, and remotely controlling the path of rockets and missiles. In this era, three prime technologies had to be improved:

- Flight power source
- Stabilization
- Navigation

BEKAA VALLEY, LEBANON

I think it is true to say that the birth of modern drones evolved out of the Israeli desert, when someone had the bright idea of flying a fixed-wing model aircraft with a camera strapped to it—this might also explain why Israel has such a grip on the UAV market. Nevertheless, it did not take long for the American intelligence world to realize that here was a novel way of obtaining information over enemy occupied territory. That's when the growth in modern unmanned war planes really began.

During the seventies, and through its checkered military history, Israel Aircraft Industries (IAI) developed into one of the world's leading drone manufacturers. During this time, the Israeli military was becoming progressively more interested in using UAVs for military purposes. The American-made Chukar target drone was used during the Israeli counterattack against Egypt in 1973. Its main aim was to get Egypt to reveal its radar sites.

One of IAI's more successful drones was the Scout, which first operated combat missions in Angola by the South African Defense Force. In 1982, the IAI Scout drone was used to combat the threat by Syrian surface-to-air missiles (SAMs) positioned in the Bekaa Valley of Lebanon. Using the same techniques and lessons Israel had developed fighting Egypt, they used these drones to flush out the Syrian SAM site locations. All 28 SAM sites in the Bekaa were destroyed. The American military was quick to learn from the Israeli battlefield success of deploying UAVs and, by1984, had embarked on a massive program. While the American military is highly reluctant to purchase its hardware from foreign countries, initially it formed cooperative companies that worked closely with counterparts in Israel. While the US has gone on to develop its own family of drones, Israel remains at the very forefront of the UAV industry.

Drone technology increased dramatically during the 1996 war in Bosnia, when Predators flew over six hundred missions. This gave way to improved command and control of operating unmanned drones in a manned aircraft environment. Today these large drones navigate the skies on the prowl for enemy locations and can deliver an instant unseen retaliatory strike if required.

Author's Note: While there is a tendency to think that military drones only fly military missions, you would be wrong. In 2012, a Predator drone patrolling the American-Canadian border spotted cattle rustlers, which led to the first ever arrest due to drones. When cops visited the farm to collect the cattle, three of Rodney Brossart's sons—Alex, Jacob, and Thomas—confronted Nelson County Sheriff Kelly Janke with rifles and shotguns and would not allow officers on their land. The Brossarts, a North Dakota family, were deemed so dangerous that the local sheriff needed an unmanned drone to provide surveillance to keep watch over the farm. During a 16-hour stand-off, the sheriff and his deputies waited until they could see the remaining Brossarts put down their weapons before arresting the three sons.

The debate about drones in both America and Europe for military purposes continues unabated, focused largely on the morality of drone warfare. Are they an unacceptable moral hazard, or are they a tool to combat terrorism? The number of innocent civilians killed (collateral damage) is a major factor and one that cannot be ignored even by the pro-drone vote.

There has been a call on the Obama administration to reduce the dependence on drones, but for the moment it seems they are the weapon of choice. Despite all the rhetoric, drones have done their job with great efficiency by successfully killing terrorist commanders in places hereto not accessible. They have penetrated deep into the terrorist refuges of Pakistan, Yemen, Somalia, and Syria, scourging organizations like al Qaeda and ISIS. They have done so without putting allied soldiers in harm's way while reducing the overall cost of warfare.

However, being on the receiving end of a drone strike is not much fun. A bomb released from a Reaper drone will damage everything within a 200-meter radius. So, imagine a football field full of people and when the bomb hits you have a 5 percent chance of living, albeit with critical injury. There are claims that drones kill thousands of innocent civilians, which sets a dangerous precedent as much of this criticism is valid.

The sophistication of the MQ-9 Reaper is really futuristic, yet here we are in 2016 with an unmanned war plane capable of delivering unprecedented destruction with great accuracy over vast distances.

At lot of work has been done to minimize the collateral damage and reduce civilian casualties. There have been a lot of improvements in re-configuring the Hellfire anti-tank missile to fit onto a Predator and Reaper drones, which is the type of missile we see taking out a moving vehicle. There has also been a change in the targeting scenario, making sure the weapons fired are equated to the target they are attacking.

Another problem, not widely published, is the number of drones that crash for one reason or another. Up until January this year, the US has lost or crashed at least 85 drones, with an average cost of around $1 million each. Some go down due to lost data-link communications or flying in extremely poor weather, while human error has also been reported. Yet neither collateral damage nor crashes have put a stop to the use of drones; if anything their use has expanded.

MQ-9 REAPER

The MQ-9 Reaper is manufactured by General Atomics Aeronautical Systems, Inc. Reapers can do just about anything: gather intelligence, carry out surveillance, and hit a designated target really hard, as the MQ-9 Reaper is an armed, multi-mission, medium-altitude, long-endurance, remotely piloted aircraft. It can carry a combination of AGM-114 Hellfire missiles, GBU-12 Paveway II, and

GBU-38 Joint Direct Attack Munitions. In addition to its many other highly robust sensors, it can employ laser-guided munitions like the Paveway II. It's known as the Find, Fix, and Finish drone.

The MQ-9 can be packed into a single container and shipped by C130 Hercules transport aircrafts anywhere it's needed. It can take off and land using local line of sight (LOS) communications and, once airborne, it switches to Predator Primary Satellite Link (PPSL) during its 1,000 nautical mile flight range. It requires two pilots: one to fly the drone and one to act as sensor operator.

Reapers are not the kind of drone you would want to crash, as they cost some $64.2 million each with sensors but excluding munitions. Between the US and the United Kingdom, there are over a hundred in use; most fly out of Creech AFB in Nevada or RAF Waddington in Lincoln, UK.

Power plant: Honeywell TPE331–10GD turboprop engine
Wingspan: 66 feet (20.1 meters)
Length: 36 feet (11 meters)
Height: 12.5 feet (3.8 meters)
Maximum takeoff weight: 10,500 pounds (4,760 kilograms)
Payload: 3,750 pounds (1,701 kilograms)
Speed: cruise speed around 230 mph (200 knots)
Range: 1,150 miles (1,000 nautical miles)
Ceiling: Up to 50,000 feet (15,240 meters)

COMMERCIAL CIVILIAN DRONES

The best civilian use of drones for the foreseeable future is where they pose the least amount of threat to human life. Rescue at sea, search and rescue in remote areas, and farming land all fall into this category because, if there is a failure and the drone falls from the sky, there is little or no possibility of it hurting anyone or causing much damage. They are also extremely useful in areas of great danger to humans, such as detecting and dealing with nuclear radiation or leaking toxic gasses.

Somewhere in between the larger military drones and the small hobbyist play-things is a whole new field of drones being developed. While the media reports military drones striking some terrorist group or some idiot flying his newfound toy near an airport, there are a lot of smaller drones hard at work doing what they do best: "dull, dirty, and dangerous" assignments. This is where the real innovation is being developed—inspired ideas to use drones to make our lives easier and more productive, or to offer assistance in the event of a disaster. Today, drones are being successfully deployed in police work, the film industry, farming, fighting wild fires, searching for victims after an earthquake, and a great deal of other uses. This "Drones for Good" crusade is growing in strength, aided by people who share a dream whereby drones save and improve lives.

Author's Note: It seems like everything about drone technology is contested: its novelty, legality, morality, utility, and future development. Notwithstanding this, last year saw an open competition for drones in Dubai call "Drones for Good." I attended the last two days and was impressed with the venue and the thought that the contestants had put into the competition. Not only was this the first event of its kind in Dubai, it was the first ever such event in the world. There were drones to dissipate fog, drones to plant trees, firefighting drones, and even a drone which could deliver a human heart from one hospital to another in order to save a life.

One of the most common uses for potential drone activity is the desire to reduce costs for industry and make things more efficient. The general perception is that drones are either military destroyers or big boy toys. This is not the case; somewhere in the middle is the drone built using advanced manufacturing techniques, with commercial off-the-shelf components. The economics of civil-use drones has been improved dramatically; consequently, this has opened up a new commercial market full of potential. Some examples:

- Fishing fleets: Monitoring of illegal fishing or fish stock location
- Environmental: Oil spills, pollution monitoring, iceberg surveillance
- Disaster relief: Governmental and charitable organizations
- Land surveying: Geo/Digital mapping
- Infrastructure security: Ports, nuclear power plants, wind farms, oil rigs
- Cross-border immigration control
- Anti-piracy
- Search and rescue including a life craft/jacket deployment
- Meteorology
- Agricultural surveying and sampling
- Communication relaying
- Forestry management

Our planet needs food, and as the population grows so does the demand on farmers. Farming is tough, and drones are making it easier. With their aerial abilities, drone can now help farmers see if their irrigation systems are working, how their crops are growing, and even detect whether any of the plants are sick by using infra-red technology. This enables farmers to make critical decisions about where and when to fertilize, plant, or water. When you couple a drone with automated farm machinery, you take the human error out of the equation. The end result: better crop yields at a better cost.

Take the film industry. The use of drones allows directors to take footage from incredible angles, achieving effects that otherwise would be done by wires, cranes, and more traditional apparatuses to a limited effect. Drones can get high above the scene, keep pace with a running man, and fly over water or clifftops. They are a key innovation in the film industry.

For the media, drones can cover a news story without getting in the way of emergency services. Drones can help firefighters by tracking the fire's movement as well as locating firemen that may have become cut off and offer them a safe escape route. By the same inference, drones can help search for survivors after an earthquake or tsunami using thermal cameras to detect body heat. Natural disasters and other types of emergencies call for timely distribution of medication and aid. Drones are ideal for this and can do it efficiently.

> **Author's Note:** Spain is no stranger to forest fires, and as such has a remarkable and rapid system for dealing with them. Several years ago a forest fire broke out near my home and, within an hour, firefighters had it under control, but asked that I keep an eye on the fire. I was requested by the local police to put a drone over the area—which I did. I think there was more interest in the drone than the fire, and everyone thought it a great idea (this was prior to the blanket ban on drone flying in Spain).

Last year, some eight hundred teams from nearly sixty countries registered for the "Drones for Good" competition, which was rounded down to five ideas that made it through to the finals. There was a great deal of innovative thinking on how to use drones for humanitarian reasons including saving lives at sea. The competition proved so popular that it is being repeated in 2016, and with a purse of $1 million, there will be no shortage of competitors.

THE MARID

Not being one to miss out on a prize of $1 million, I entered my own team this year. We have developed a heavy lift drone capable of operating at sea or on land; it's called the Marid. Its main function is lifesaving, damage assessment, and delivery of emergency equipment or supplies. Six prototypes were originally built and given to MIT in Boston for its Waterfly project. These drones were flown at the 2015 Drones for Good by MIT, using its own electronics. The new Marid is three times larger and capable of carrying and delivering the newly developed Seapod. The Marid has a flight duration of 40-plus minutes, with super long range, and is capable of automatic search patterns.

The Seapod is a totally new innovation, capable of saving lives among those in danger of drowning at sea, such as those escaping the conflict in Syria and

EARLY PROTOTYPE

Commercial drones not only search for survivors quicker than humans, but can also help rescue and protect them. The Seapod can be dropped from a drone so that even small children can scramble onboard.

seeking a better life in Europe. Once dropped from the Marid, the Seapod automatically deploys when it touches the water. The design is advanced, allowing even small children to crawl onto it. The Seapod can accommodate up to eight people and carries lifesaver equipment to produce water and sustain life for up to four weeks. It also has an emergency location beacon.

The Marid can also be used over land, responding to a natural disaster scenario. The drone can be a source of valuable information by carrying out an immediate damage assessment, while at the same time searching for survivors. Its secondary role is to aid those survivors by providing emergency supplies to inaccessible places.

An early damage assessment is one of the most important factors in responding to an earthquake, typhoon flooding, or a tsunami. It takes 875 man hours to search one square mile, whereas an autonomous drone can do it in 40 minutes. Using thermal cameras a drone can detect heat signatures in rubble and get medical aid and survival equipment to those that need it rapidly. Best of all, a drone can work in any dull, dirty, or dangerous environment.

The drone itself is a VTOL (Vertical Take Off and Landing) which presents a new standard in aerodynamics, flight stability, payload capacity, and flight time. It is designed to operate in high winds and adverse weather (35 kmh wind burst) while carrying a massive 10-kilo payload, using an ultra-light carbon gyro-sta-bilized gimbal with visible and thermal infrared cameras that can be utilized both day and night.

Using the new UX400 purpose-designed Ground Control System, the Marid can be safely flown after a few hours' instruction, although a week's course is recommended. The UX400 not only allows for rapid autonomous flying, but also outputs the video downlink to any authorized person worldwide. As it provides a vast amount of exceptional visual information, it prevents collision accidents by scanning the skies for other aircraft at a distance of up to 200 miles.

Marid Features and Capabilities
- Capable of deploying the new self-inflating seapod
- Highly stable auto hovering with medical, location, and survival package drop
- Autonomous navigation—way-points and search pattern based missions
- Automatic takeoff and landing

The mighty Madrid is a variation of the UAVision Spyro, which is manufactured in Portugal. It is one of the largest octo-copters in the world and can be used for a wide variety of tasks, from search and rescue to helping farmers produce better crops more efficiently. This brilliant drone is currently in use around the world.

- Bidirectional data link with 7½ mile (12 km) range
- Auto stabilized camera payload (pan, tilt, and roll gimbal)
- Digital encoded video link for worlwide distribution
- Visible, thermal, and NIR cameras
- Automatic target lock system
- Advanced flight and NAV control system
- Autopilot black box

By hobbyist drone, I am referring to all those we now see in the stores or online—the ones that your significant other or child wants for Christmas. A top FAA official predicted that as many as one million toy drones could be sold during the 2015 Christmas holiday. It's not just the US that's under drone attack: in the UK almost every major outlet is offering at least ten different varieties. Drones have indeed caught the imagination of many. They range in price from $100 to $4,000 at the top end, and although the cheaper end will simply fly—albeit in a rather ungainly fashion—the top end offers extremely smooth flight control, with functionality and a video downlink quality even the military would find difficulty to match.

Trust me when I say that some of the newest drones are incredibly sophisticated—not just in their functionality but also in the aerial imagery they can produce. Many now use 4K camera systems with boosted communications out to a range of 3 miles (5 kilometers). Given the average size of a hobbyist drone, that is way out of human sight, therefore control is via the camera and the telemetry

While most drone sales will be of the smaller cheaper "toy" models at the other end of the spectrum, the DJI Phantom 3 and the 3DR Solo can be purchased for around the $1,000 mark—and I can assure you these are extremely professional drones.

mapping. I have been in this industry for many years, and the ingenuity never ceases to amaze me. From functions such as creating a search pattern that the drone will follow to having it circle an object and climb up the object as it does so, the ingenuity still blows me away.

DRONE MANUFACTURING

Globally, the total drone market is expected to more than quadruple to be worth over $4 trillion in the next eight years. This market is driven and currently dominated by military and commercial users, but the hobbyist market is closing rapidly. The current market in the US is worth $36 billion; in Western Europe, it is worth $9.9 billion; and in China, $9.8 billion. Over a third of the drones in the world are made in the US. The second-largest known manufacturer is Israel, with approximately 16 percent of the market share. Many of the Chinese and Russian military drone programs remain secret.

In the US, it is anticipated that the US Department of Defense will nearly double the number of unmanned aircraft in its fleets by 2021. Meanwhile, the use of

The staff of BCB Robotics at EU headquarters demonstrating a model of the seagoing border protection drone, similar to that supplied to MIT.

drones is increasingly becoming accepted doctrine within the European armed forces and civilian authorities alike, in a context of increased sensitivity to risking human life combined with cost-efficiency requirements. Drones have demonstrated their benefits, advantages, and strategic potential that could constitute a valuable asset for the requirements of European missions.

Although remotely piloted drones can perform tasks that manned systems would not be able to perform and go on missions where the lives of the pilots would be put at risk, there are some considerations that need to be addressed. These include:

- Lack of airspace regulation that covers all types of drone systems
- Liability for civil operation
- Capacity for payload flexibility (the heavier the drone, the more damage on crashing)
- Lack of secure nonmilitary frequency for civil operation
- Perceived reliability (vehicle attrition rate versus manned aircraft)
- Operator training issues (who needs permission to fly what)
- Recognition/customer perception (Drones for Good)
- Safety (drone safety record must be higher than manned aircraft)

DRONE TECHNOLOGY

In the years I have been flying, there has been a massive shift in technology. True, in the seventies I had a radio-controlled plane which I could fly around 200 meters.

The Hycopter is an "H" configured quad-copter that is currently being tested using ultra-light fuel cells created by Horizon Energy Systems. It is hoped that the drone will fly for at least 2.5 hours while carrying a payload of 2 pounds; if this works it will be a massive leap forward in flight duration.

However, in the past eight years since drones entered the arena, there has been a massive leap in technology, especially in the commercial and hobbyist markets.

Using new materials for the airframe, improved communications, and better propulsion systems, drone technology has improved dramatically. One of the biggest physical restraints on smaller drones is power supply. Batteries can only hold so much energy, and because adding more batteries to a drone also increases the weight of that drone, there are finite limits on how long quad-copters can fly in a single flight. Most of the smaller commercial drones are powered by lithium polymer batteries, but it seems this technology has reached its peak and a new energy source is needed. Hydrogen fuel cells may be the answer. Fuel cells, which create an electrical current when they convert hydrogen and oxygen into water, are attractive as energy sources because of their high energy density. Tests using hydrogen fuel cells to power fixed-wing aircraft have been highly successful, resulting in a dramatic leap in flight duration.

Possibly the hardest technological challenge to meet is the need for long-range communications between the pilot and the drone. Traditional frequencies for this task have been pushed to the limit, and while good ranges have been established for radio control and telemetry, the bandwidth required for transmitting video remains an issue.

Although the military has dedicated satellite communications for the control and transfer of data on its larger drones, the rest of the industry is left to rely on ground communications. However, there is a shift in thinking taking place. The problem is the type of communication being used to fly the drone and the length of time required for instructions and data to be passed between the ground control station and the drone. Latency and missed messages can cause the drone to fly off course or do things it should not. Long-range communications are improving rapidly, and no doubt some bright engineer will develop a solution.

Author's Note: At this stage, I will offer you a word of warning. Just because you have designed, paid for, and built your drone does not necessarily mean it's yours to sell. What do I mean by that? It's simple: there is a law called ITAR (International Traffic in Arms Regulations). Although most people think this law only applies to the military export of drones, that is wrong. For example, if you build a drone that has some special features you developed yourself that could be of military use, and if you used American parts or technology to make it, it can be construed as ITAR and therefore restricted. Trust me, I know. Thus, if you have developed a drone with unique capabilities, then you are seriously advised to check out the law on selling it domestically, and certainly if you intend to export it. To ignore this advice could mean a heavy fine or possibly a custodial sentence. In particular, anything that you build to drop from a drone with an internal guidance system also comes under the ITAR law.

SUMMARY

The rise in the usage of drones and the numbers being manufactured clearly indicate they are here to stay. Although there can be no doubt as to drones' effectiveness within current society, there are still many issues to be addressed for all types of drone.

Should we use military grade drones to monitor our borders? Should we try and weaponize many of the smaller drones for use in war? The answers to these questions are mainly out of our hands; but the use of drones for peaceful purposes is certainly within our reach. In the latter case, we are only barred by law concerning where we can fly and the restrictions placed upon us.

The path of parcel delivery by drone is subject not just to the "what if" factor previously mentioned, but to weather conditions. For areas of low human population and where human life is at risk, commercial drones will most certainly come into their own. As technology and safety advance, so will the functionality and usage of drones.

Chapter 2

READY MADE OR BUILD?

As a child I would often spend my hard-earned pocket money on a Keil Kraft balsa wood model, which I would build on a Saturday evening. Sadly, come Monday there would be very little to show for my efforts. Keil Kraft were mostly powered by an elastic band which ran the length of the aircraft from the nose cone to the tail internally. Sometimes I would wind the elastic up so tight that the tail section would end up in the nose cone as the frail body gave way to the pressure. They were hand launched and free flight with little or no control at all—but what fun they provided!

My first real RC build did not happen until I rekindled my childhood hobby at the age of twenty-five, by which time I was a soldier and a member of the British Special Forces. It took me months to build the plane and fit the radio control unit, which was primitive by modern standards. Finished, I took my pride and joy to the local park. With full throttle and a hefty throw, it dropped straight to the ground— it would not fly. I had ignored the warnings about making the aircraft too heavy. It was hard lesson to learn, but then I was always a perfectionist when it came to building solidly.

I left it for some months but then, one fine afternoon, I took it to the military barracks with me and ran it up and down the soccer field. I was happy that the

My very first model aircraft that took me ages to build, only to lose it on its maiden flight.

motor started and that I had some control via the RC handset. But it just would not take off. I next tried hand launching the plane under full throttle; it still dropped to the ground, albeit gradually. So, in one final effort I just let it run at full throttle along the ground. After some 165 feet (50 meters) or so it started to bounce a little, so I kept the throttle at full. Then it lifted gracefully into the air—happiness. With it flying at little more than shoulder height, I ran after my plane only to see it climb out of reach. I cut the throttle, but the plane continued to fly on. Throttle full down and still no response. My first model plane climbed to a height of around 165 feet before turning to the left and heading into the nearby town. Aviation law was quite lax in those days and, though I was not worried about it hitting anything, I wanted my plane back. I drove my car in the direction it was last seen heading and, would you believe me, after a short search I spotted a group of women and children as I drove down the street—they were holding my plane. It had landed perfectly on a garden lawn about a mile away from the launch site with no damage whatsoever. Second lesson learned: get myself properly trained, and I did.

Many years have passed since then; things have changed, and once more I find myself involved with flying machines . . . well, drones to be more precise. I

One of the first drones produced by Middlesex University. It flew extremely well, did all that was required, and prompted me to engage in some serious drone development.

had been developing a tracking system for Special Forces when the thought came to me that it might prove prudent to also have an "eye in the sky," and so I went in search of my first drone.

The best I could find at the time was a Microdrone from Germany, but it cost close to 20,000 euros. That was nine years ago. And although the Microdrone was good, it was not exactly what I wanted. So I immersed myself in drone technology. I thought about designing my own, but although my knowledge of electronics is not bad, there was no way I could build a modern drone from scratch. I decided to enlist the help of someone who could. This turned out to be Dr. Steven Prior.

Dr. Steven Prior ran a small unit at Middlesex University that dedicated itself primarily to multi-rotor aircraft and, together with a small team of students, built a range of excellent drones. He was project lead for the UK MoD Grand Challenge i-Spy team in 2008, and also the leader of Team HALO, winner of the DARPA UAVForge Competition 2012 in the US. Steven has since moved to Southampton University, where he has set up a similar project, concentrating I believe on tethered drones.

In 2011, I was introduced to Ivan Reedman of Torquing Technology Ltd. in South Wales, UK. Ivan was a self-taught genius who offered to help develop a

The first of the SQ-4 series was far too ambitious in its undertaking and complexity.

drone for our company. To make a long story short, we progressed into the darkness of unknown electronics but, if I am truthful, we were far too ambitious. Our first drone had total sense-and-avoid capacity using six sets of sonar plus a whole host of other advanced features, such as voice commands for control. The first SQ-4 was born, and it did fly, albeit not far. Looking back, I think the main problem we encountered was the communications; we used Wi-Fi, and it just did not have the range, despite pushing the power way beyond the safety limit.

While I took another path and went back to the drawing board in pursuit of my original goal, Ivan pressed on with his own ideas. In early 2015, Ivan launched a Kickstarter program that netted his company £2.3 million ($3.5 million) for his Zano program, the fastest ever fundraiser in Europe. He had used all his knowledge to develop a small short-range drone which was ideal and perfect for Wi-Fi. It was to be a highly advanced "selfie" drone capable of sense-and-avoid, as well as swarming. Unfortunately, the project failed, and the company went into liquidation, leaving a lot of angry Kickstarter backers.

My new path led me to look back at all the developing and different methods of drone control. In the end, I discovered a company in Portugal, UAVision, which had just what I wanted, and shared my ideas and aspirations on drone development. The UAVision system followed the APM and Mission Planner route, but with a lot of modifications as the company used its own code and developed its own autopilot. I continue to work with them and consider myself privileged to do so.

As with most things, it's a matter of progressing to see what works and what does not. In the arena of drones, we find a vast variety of imitators—those using other people's work and innovation. Yet from time to time we also find the odd

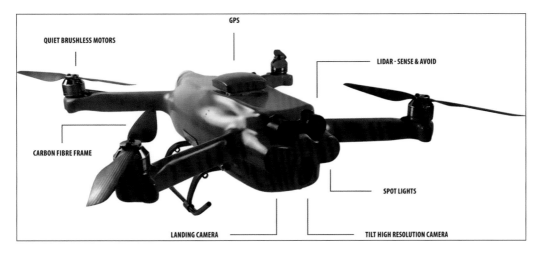

GPS
QUIET BRUSHLESS MOTORS
LIDAR - SENSE & AVOID
CARBON FIBRE FRAME
SPOT LIGHTS
LANDING CAMERA
TILT HIGH RESOLUTION CAMERA

The very latest SQ-4. A small but brilliant drone designed primarily for search and rescue but with both military and police work in mind. It took several years to get it right.

genius who comes up with something entirely new. Sven Jurgens of Microdrone spent years developing the early quad-copter with his father; Doctor Steven Prior has spent years researching drones and offering valuable advice, while Ivan Reedman proved that you don't have to have a degree in electronics to fulfill your dream. The team at UAVision in Portugal develop from the concept up, through the design phase, to final production. In doing so they research, learn, innovate, and advance drone technology at the same time.

YOUR CHOICE

The choice of building your own drone or simply buying one is strictly up to you. If you have the money, you can purchase a drone today and have it ready to fly within a few minutes. If you're a complete novice, the odds are you would certainly crash it, or lose it before the day is out. Even a new drone needs some care and attention, such as checking that it's calibrated properly and ensuring the battery is charged. Moreover, you need to understand how it flies and learn some extremely important safety rules.

Unlike a fixed-wing drone, a quad-copter is a flying brick. It has literally no aerodynamics about it whatsoever; throw it without the motors running and it simply falls to the ground. It relies totally on its electronics to keep it flying in a stable mode. Yet when it does fly it can be made to do so extremely simply—and that is the wonder of any multi-rotor drone. It takes off vertically and hovers in the air waiting for commands, and that is its one exceptional feature over fixed-wing. For me, the most important lesson to learn is that quad-copters float on air, which means they are subject to two main influences: inertia and wind.

When your drone is flying in any direction, it will generate with it a certain amount of inertia. The amount generated will be dictated by the drone's weight and speed. This means that when you are flying and decide to stop, there will be some continuous movement of the drone. True, this can be corrected by using the joysticks, but first-time flyers often overlook inertia. Owing to the fact that your drone is simply floating in the air, the effects of wind also make a difference. Again, this can be corrected by using GPS and the joysticks. Additionally, when you fly very close to a building or wall, the drone tends to get sucked into it due to the down-wash of the propellers. If you ask any "newbie" flyer what was the cause of their first accident, it will likely be one or a combination of the above.

Having a drone is only part of the equation; the other part is learning to fly safely. Provided you have a half-decent autopilot, you can get anything to fly. My friend Tim Whitcombe, a superb RC pilot, surprised me one day when he turned up with a quad-copter made from electrical tubing he had purchased from the local DIY store. It flew. With no GPS and very little configuration, and without all the functions we take for granted today, it flew. Other than the motors and propellers, Tim had made it from scraps he had left over from his other models, and put it together in only a few hours. I should point out that Tim Whitcombe has years of experience building and flying model aircraft from very small helicopters to extremely large jet aircraft. I *do not* recommend that you start your build off by making your own frame, as this is almost certain to end in disaster.

MULTI-ROTOR OR FIXED-WING?

It is also important to know where our drone fits into the scope of avionics. We are all used to seeing military drones flying in a similar manner as a normal commercial aircraft with fixed-wing. Until a few years ago the bulk of the hobby market was also fixed-wing (and a few helicopters). Yet when we look at the commercial market today, we find that most of the nonmilitary drones are in fact multi-rotor. Multi-rotor means a drone with either four, six, or eight arms or motors, of which the predominant model is quad-copter with just four arms. While we are concentrating on building a quad-copter, it is worth

Vertical takeoff and landing drones come in all shapes and sizes. This drone can map using sonar.

learning a little about the difference between the multi-rotor and the fixed-wing.

> **Author's Note:** When learning about drones it is easy to become confused about the various shapes. For example, you can have a tricopter with three arms, a quad-copter with four arms, a hexacopter with six arms, and an octo-copter with eight arms. But it does not stop there. If you take a tricopter and add three more motors (top and bottom of each arm) you get a Y6 tricopter, and the same goes for adding four more motors to a quad-copter making it an X8 octo-copter. The main reason for having more than four arms is redundancy; if one motor fails, then a hexacopter or octo-copter will survive better than a quad-copter, which needs four motors to stay airborne. In addition, the more arms you have, the more power and lift, but also there is a greater demand on power supply.

Multi-rotor

There are several major advantages of multi-rotor over fixed-wing drones and small helicopters. They take off and land vertically (VTOL), they hover in the same spot, they do not have all the mechanical linkages of a helicopter, and, in the case of a quad-copter, the four motors allow for a smaller diameter than that used by a helicopter. Likewise, there are some limitations to a multi-rotor, such as range and duration. However, new power supply technologies are expanding, minimizing this disadvantage.

The drone's enhanced functionality is its capacity to hover precisely over a given point and the ability to carry out automated tasks or be manually operated while stationary. In addition, drones have a low noise signature, which is essential for operations with a wildlife, agricultural, or urban surveillance role. Finally, they require no runway as they can land in extremely small spaces, or even perch on the lip of a building and use their camera systems to observe.

Fixed-Wing

RC fixed-wing aircraft have been around for a long time and, despite the rapid growth of multi-rotor drones, remain the backbone of the military and aero-model industry within most flying clubs. While there has been a lot of advancement in fixed-wing aircraft over the years, the basics of flight have remained the same. For example, aerodynamics still provide lift by air passing over the control surfaces.

Control of the fixed-wing drone is built into the wings, which traditionally consist of ailerons, elevator, and rudder. They allow the fixed-wing drone to rotate freely around three axes that are perpendicular to each other and intersect at the drone's center of gravity. The elevator controls the pitch (lateral axis), ailerons control the roll (longitudinal axis), and the rudder controls the yaw (vertical axis). This airspeed is generated by means of a propeller.

A lot more skill is required when you fly fixed-wing aircraft, and a lot more precise judgment is required. Smaller models can be hand launched, or if the ground is suitable they can take off like real aircraft, using a runway. While in flight, learning to turn the aircraft without crashing it takes a lot of skill, but the real test comes when you try to land it. A fixed-wing aircraft needs to be lined up into the wind with its landing site and be at the right approach speed and height. If the aircraft is coming towards you and you are not experienced, you can easily get confused and crash either before or after your landing site. It takes time and a lot of practice to safely fly a fixed-wing aircraft.

Military and Commercial Fixed-Wing Drone

There are many reasons why fixed-wing aircraft remain in demand throughout the commercial market, with flight duration and range being the principle motivators. A fixed-wing drone consists of a much simpler structure in relation to a

Hand-launched fixed-wing.

Fixed-wing aircraft launched from a small catapult ramp.

multi-copter, added to which it is aerodynamic, meaning the power-to-weight ratio is much lower. The simpler structure provides a less-complicated maintenance and repair process, thus allowing the user more operational time at a lower cost. Finally, in many cases, the flight characteristics are such that the drone retains some natural gliding capabilities when power is lost.

Naturally, the larger military drones are launched and landed from an airstrip in the same way as a passenger carrier aircraft. However, many of the smaller fixed-wing aircraft used by both the military and commercial market can be launched by hand. Recovery will depend on the model size and weight, but in the main those weighing less than 8 to 10 kilos (17 to 22 pounds) can be safely recovered on the ground. Larger models will require a catch-line, parachute, or net for recovery.

A popular method of launching both military and commercial aircraft is the catapult launcher. These are partially good at ensuring a clean liftoff where space is limited—for example, launching a drone from a ship. The Boeing Insitu ScanEagle is possibly the best surveillance drone ever made, and is a perfect example of fixed-wing aircraft abilities. It has flown numerous military missions, launched and recovered from both land and ship by catapult.

Hobbyist Fixed-Wing

As previously mentioned, enthusiasts have been building model aircrafts and flying them for many years. From the start—and until recently—the majority of these were fixed-wing aircraft, and that remains true to this day. The models built ranged from copies of real aircraft to those designed by their owners, both big and small. The hobby grew into an industry that spans the globe, and today we see a mixture of original models together with jets powered by a real miniature jet engines capable of over 440 mph. In 2009 a dynamic soaring glider flown in Weldon, California, reached speeds close to 400 mph. Both of these feats were achieved by enthusiasts with years of flying experience, but even they had to start somewhere.

Author's Note: I first learned to fly with fixed-wing model aircraft, and I think it really helps when it comes to flying quad-copters. It is true that just about anyone can fly a quad-copter right out of the box, but in doing so there is a lack of understanding and responsibility. Learning to fly with fixed-wing requires not only skill, it introduces discipline and responsibility not found with the advent of multi-rotor craft. Those who have learned to fly with fixed-wing will likely be much better drone pilots than those who have not.

High-Wing Trainer

If you do decide to go with a fixed-wing model first, you are advised to select your trainer with care. There is no doubt that a *high-wing*. model aircraft is much

FMS FHX 1280.

easier to fly than, say, a model Spitfire, which is a *low-wing*. Moreover, in the case of a fixed-wing aircraft, you are better off purchasing an RTF model, one that you can get into the air with minimum difficulty. Within this range, you will find both 3- and 4-channel RC planes being offered—I personally would go for the 4-channel, which offers you control over the motor, ailerons, elevators, and rudder. I say this because I have found that the 3-channel reduces the learning curve.

It will also help if you choose a model that is a slow flyer, and you will soon learn to love it, not for its looks, but for its performance. Similarly, you are better off with an electric model that is powered by a battery rather than a combustion fuel-type engine. You are going to crash it, I promise, therefore choose a model that is made from polystyrene. These models are lightweight and fly really well, and tend not to get too damaged when they crash. Polystyrene is also much easier to repair than balsa wood.

Finally, do not buy a model that is either too big or too small—too small and it's subject to wind, too big and it means more damage when it crashes.

I am not saying a high-wing trainer is the best fixed-wing aircraft when you want to learn, but it does check off most of the boxes. Not having flown a fixed-wing for several years, I purchased an FHX 1280 and was pleasantly surprised at the performance. It has a 50 1/2-inch (1,280 mm) wingspan with full 4-channel operability. The flying is really smooth, and it provides a great learning platform. If there were any drawbacks, I thought the motor was a little underpowered, especially when flying into strong wind.

THE DECISION TO BUY OR BUILD

The best thing to do is gauge your experience and skills for building a quad-copter. Next, decide how much time you can commit to the hobby. Finally, what do you want from your quad-copter? If your end goal is simply to make something that flies, then I would say build from scratch. If your time is limited, go for a kit solution. If you aim to become a professional aerial photographer, it should be something in the middle. Whichever path you choose, build slowly and build while understanding. Even if you intend to fly using telemetry, or place a camera on your drone for first person view (FPV), just build a basic drone first while using a traditional radio control unit to fly.

I am assuming by now you want a drone, so let's discover the best solution for you. There are a lot of options, and because of the cost you should consider all of them carefully. In recent years the price of commercial drones may seem to have dropped, but personally I know this is not correct. It just seems that way because many Asian manufacturers have jumped on the drone bandwagon and made their own versions. Personally, I would not recommend you waste your money on a cheap drone. The more serious and best-marketed drones have in fact gone up in price due to the increasingly complexities and functional abilities of the models.

SECONDHAND

Buying secondhand is another solution, but I would caution you to be careful if you intend to go down this route. Don't get me wrong, there are a lot of financial benefits to buying a used drone, and there are some great deals out there. The first thing to ask yourself is what you want to spend and then what can you get for your money? Finally, is the drone you intend to purchase in good working order?

You will find a lot of people upgrade their drones as new models come out, and to help finance this they sell off their older models. Do not be afraid of buying secondhand; for example, a DJI Phantom 2 is a great drone, as is the 3DR Iris. Both are perfect first-time drones. Another perk of buying secondhand is the amount of extra equipment you might pick up, such as extra batteries, and so on. These can be expensive items to buy separately.

So the best advice is to check out all the advertisements for used drones and see what is being offered. Personally, I would stick with a brand name and buy one that is complete, meaning that it includes the controller, spare propellers, battery, and so on. Getting a camera thrown into the deal is even better. If it's in

There is no doubt about it: the DJI Phantom 2 with its own built-in camera is a gem of a first drone. There are a lot for sale on the open market as owners upgrade to newer models. A quick search on the Internet found some as cheap as $450 (£300). The Phantom 2 has most of the things you need when learning to fly a quad-copter—it even has some way-point flying features using an iPhone.

a specially designed case, the chances are it's been well looked after. I find its best to buy locally if you can, so you can go and take a look and see the drone actually fly. It is always best to take someone with you who knows about drones and who can test fly any potential purchase. If the seller is local and a good pilot, get them to give you a few tips and a quick flight lesson for free.

When searching for your drone, you will come across many strange acronyms. In fact, the drone world is full of them. Don't worry—all will be explained throughout this book.

DIFFERENT TYPES OF PACKAGES

RTF

Ready-to-fly drones also have the appeal of working right out of the box; you are not required to purchase any further components. Typically an RTF drone will have a controller or an app controller that you can download free of charge. Depending on the size of the drone purchased, you may be required to fit the propellers and of course charge the battery. While this is possibly the most expensive method, it will be all-inclusive. Additionally, there is a good chance it will fly fairly well with predicable performance.

BNF

Bind-and-fly simply means you get a complete drone as with the RTF, but without the controller. With many BNF drones, the controller is usually a handheld RC device. Basically, it assumes you are already flying and have an RC controller that is compatible with the BNF drone you intend to purchase. This is a cheaper and better option, plus you will be familiar with your RC.

Author's Note: Do not automatically assume that just because you have an RC control it will work with your purchase. Many bottom-end "toy" drones come with what looks like a real RC device, but normally these will only fly the unit for which they were designed. You need a top-end hobby RC that has a wide variety of functions and controls, making it compatible with a range of different RC receivers.

ARF

Almost-ready-to-fly can mean a lot of things, but usually it means a kit form model that requires assembly. ARF may or may not be equipped with a controller, but it should arrive with all the components. In essence, it is a DIY job that often requires using your own electronics. It will also require some building skills and knowledge of quad-copter setup. The end results will most likely not be as smooth as those of an RTF model.

KIT FORM

This will mean a lot of work and possibly more time before you get your drone in the air. It will definitely require good building skills, and the end flying results may vary. On the upside, it should prove to be the cheapest option (though not always), and you will learn a lot of basic skills. Most of all, it will be fun.

WHAT TO BUY?

Today there are so many drones that it's almost impossible to advise where to start. Many drones you find in model shops or for sale online are produced in China. Some are extremely good, and some are total rubbish—be careful and do your research. Go online and look at YouTube videos of drones being unpacked, flown, and reviewed. You will hear and see a lot of names come up continually: 3DR, DJI, Walkera, Hubsan, GoPro, FPV, FatShark, and many others. This is because these brands have been proven to be of good quality with a high degree of customer satisfaction. I believe that nothing under $100 flies or lasts long enough to satisfy the basic desires of drone flying. So let's look at some of the better models.

DJI

DJI is a Chinese company operating out of Shenzhen, Guangdong province. It was started by Frank Wang and has developed into the leading drone manufacturer worldwide. I own four DJI Phantoms, starting with the original, two Phantom 2s, and a Phantom 3. They fly beautifully straight out of the box. As a first purchase, this has to be the best drone to go for. The reason I say this is because of its simplicity of operation and its smooth flight capabilities. What I also like about DJI is how it has continued to develop and improve its drones without losing the original features. Over the last few years, DJI has come to dominate the small drone market. The key to its success has been to make a complicated aerial robot affordable and usable by a complete novice.

With advancement comes a rise in price, but the technology in the Phantom 3 is simply amazing. It has a massive 1 1/4 (2 kilometer) transmission range and can be fitted with a 4K camera. If there is anything lacking in the DJI range, it's flight time, but the battery charge time has been vastly improved with the aid of a new powerful charger.

The New DJI Inspire

This is the latest offering from DJI, and it is a real professional-level drone, perfect for both video and still photography. It is super stable in flight, has a Follow

Me function, and standard way-point operations, making it almost as advanced as most military small drones. In fact, several police forces in the UK have already purchased the DJI Inspire for testing in various law enforcement roles.

3DR

3D Robotics is a company that has a great pedigree because the two owners, Chris Anderson and Jordi Muñoz, who met through the online DIY Drones community, really know their stuff. Anderson was an entrepreneur, while Muñoz had the skill and innovation. Today their company has expanded to become one of the best recreational drone manufacturers in the world.

What first caught my attention about 3DR was the ability of its autopilot to be controlled by Mission Planner and later DroidPlanner (both of which are covered later in this book). I purchased one of the first IRIS drones, which was released in September 2014, and I still enjoy flying it to this day. Compared to the DJI Phantom, I found I had to do more to get my IRIS in the air. It took a little more skill, which greatly improved my flying. I was always aware that I was flying it as opposed to the drone being in charge (if that makes sense). The IRIS coupled with Mission Planner seemed to be just what I was looking for and, in truth, set me on my present path of development.

In many respects, the IRIS flew better than the DJI Phantom. Though it lacked the simplicity of flight and smoothness of the Phantom, the IRIS had a great turn of speed and range. Plus flying the IRIS with Mission Planner or DroidPlanner gave the operator a wide range of features. One of its endearing qualities was that it always seemed to do exactly what I wanted it to do. Apart from using the GCS software, I also flew the IRIS with a Spectrum DX6 transmitter when I wanted a bit of fun.

Apart from producing drones, 3DR also manufactures flight components, including the Pixhawk, possibly the best and most widely used autopilot in the world. This autopilot has been fitted not just to quad-copters but to a whole range of radio-controlled devices, including fixed-wing aircrafts, ground bots, and water-related robots. We will be using the Pixhawk to build our drone later in this book.

Solo

Solo is the latest drone from 3DR and claims to be the smartest drone ever. It has two computers onboard and like its rival, the DJI Inspire, it has been manufactured to operate as simply as possible.

WALKERA

Walkera, founded in 1994, originally manufactured model helicopters. The company, which employs around 1,500 people, is located in the Panyu District of

Guangzhou, China. I purchased my first Walkera some years back, and although I had fun, it did not last long. Nevertheless, Walkera has continued to improve, and a recent purchase of the Scout X4 with its long-range F12E transmitter has proven it to be one of my preferred drones when it comes to doing some serious flying.

Walkera Scout X4

When I first purchased the Scout X4, I, was a little taken aback by its size and weight. Nevertheless, the construction was excellent and it had the look of a drone with a purpose. Initial calibration was fairly straightforward and involved a similar procedure to other drones. When it came to flying, I became preoccupied with its ability to fly from a tablet as well as from an RC transmitter. Naturally, I first test flew the Scout X4 with the Devo F12E RC controller that came with it and found that it was best to give the drone a wide berth until it was in the air. Once satisfied, I flew it using a NEXUS 7 tablet with a telemetry unit attached via a short USB cable. Walkera lets you download the GCS software for free, and I found it very similar to DroidPlanner though a lot more complicated. Auto take-off, way-point flying, and landing all worked fine, and I was impressed. Then I used the on-screen joysticks; having selected mode 2 (left throttle), the drone took off with no problem and responded brilliantly to the on-screen joystick commands—with one fault. Unlike an RC transmitter, where you actually hold the joystick, with the tablet you simply slide your finger. Removing your finger from the tablet allows the on-screen stick to suddenly self-center, causing the drone

The Walkera Scout X4, one of my favorite drones.

to react violently to the change. Once you are aware of this variance, it is easy to control and a really advanced drone to fly.

In truth, there are so many drones to choose from today, and although the leaders remain DJI and 3DR, companies such as Walkera and Hubsan have many years of experience. If you type "best drone to buy 2016" into your Internet browser and click some images, you will see what I mean.

BUILD YOUR OWN DRONE

Here you have two choices: purchase a drone in kit form or buy the individual parts to assemble one. The kit obviously has the advantage of everything you need arriving in one package; in most cases, the motors, props, and ESCs will be matched to provide the best performance. However, I am not an avid lover of kit form projects, because you do not always get the best quality components. Many kit form drones have all the parts required, but in many cases they are of inferior quality. You must remember that you are building a drone that has absolutely no aerodynamics whatsoever; it stays in the air by using some pretty advanced electronics and clever software.

Personally, I like to select my own parts for assembly, because this drives me to get a better working knowledge of each component. Additionally, should you have an accident, it is easier to purchase replacement parts and make repairs. Crashing a drone that you have purchased can sometimes mean a total replacement or sending it back to the manufacture for repair. Plus, if you've already done the research on these parts and personally installed them, you'll have a better understanding of how to repair or replace.

You should also consider what you will use your drone for. I would guess in the majority of cases it is for pure pleasure. Hopefully, some readers may have more ambitious ideas—FPV or aerial photography, for example. There is a tendency to think that drones are for surveillance and that they are new technology and a bit unsafe. I have written about two other types: a racing drone and a tethered drone. While both are in their infancy, they are a clear example of how drone technology can be used or modified for a specific purpose.

SUMMARY

Before you invest a lot of money in a drone, first look at all the options mentioned in this chapter. Consider how deep you want to delve into the subject and whether you're after fun flying or want to take a more professional approach. Do as much online research into drones as you can, especially of the reviews and YouTube videos of drones flying. Do not believe all the manufacturers' hype on what their drones will do.

I would advise those who are really serious to join DIY Drones (www.diydrones. com) or other drone-related websites, as this will give you plenty of food for thought. Don't be shy when it comes to asking questions—we all have to start somewhere! Don't let the technical jargon put you off; most of it is not required in the early stages, and as you progress you will find it easier to understand. The Glossary contains a list of useful acronyms relating to drone technology at the end of this book.

Don't just purchase a drone because of the brand name; there are also many popular individual drones, and as with everything, you get what you pay for. I found a guy in London who made a great basic drone that we used for testing more advanced components. Despite two years of use and the occasional crash, it is still in one piece. Drone technology is improving, and for the most part almost all commercial drones over $200 fly reasonably well. But at the end of the day, it is all down to price. The more you pay, the better the drone will fly for you, and the more functionality you will get.

Chapter 3

SKILLS AND UNDERSTANDING THE DRONE BUILD

First off, it is best to understand what is needed and how it all fits together when you decide to build a drone for the first time. Secondly, you need the skills to actually assemble the various components. While many are simply "push fit" connectors that fit into a corresponding port, others need soldering. If you have decided to go along with my build guide, then you should jump in with both feet. That said, it is best to progress slowly and visualize with a degree of understanding what it is you are making. In order to begin, we need to harvest our components and understand what each part plays in the overall construction.

Typical X Frame Drones.

Our drone is a quad-copter, which means it will have four motors complete with propellers attached to an X frame. This will be stabilized and answer our control demands through an autopilot that will sit in the center of our X configuration. Naturally, we will need to add power in the form of a battery to fly our drone, and that's about it.

Well . . . not quite. We will also need to control the speed of the motors in order to get proper direction. Likewise, we also need to have a radio transmitter and receiver in order to send our control messages to the drone. Then we should be up and flying. It seems simple and it is, but we can improve our drone by adding extra hardware such as GPS and telemetry.

But we are getting ahead of ourselves; let's look at all the component parts we need and what role they play.

ACRONYMS

As previously mentioned, there are a number of acronyms that you will see constantly throughout this book, and it is important that you at least have a basic understanding of what they represent in our build. (Some are explained as we progress, and others can be found in the Glossary.)

- **APM:** ArduPilotMega autopilot electronics, which many refer to as the autopilot.
- **ArduCopter:** This the software for the APM and Pixhawk electronics.
- **ESC:** Electronic Speed Control. The drone stays in the air because of the speed of the propellers (lift) provided by the motors. The motor speed is varied by the ESC depending on either the instructions from the autopilot or the pilot.

- **GHz and MHz:** Most civilian and hobbyist drones are controlled by radio signals for flight control, telemetry, and video transmission. The most common frequencies are 900 MHz, 1.2 GHz, 2.4 GHz, and 5.8 GHz. 433 MHz or 869 MHz are used for long-range UHF control systems.
- **FPV:** First person view. This is where a camera has been placed in the drone and fitted with a video downlink to either a screen attached to the pilot's RC controller or video goggles. Basically, it means the pilot has a live cockpit view from the drone.
- **GCS:** Ground Control Station/System. Software running on a computer or tablet that receives and displays information from the drone, such as video and telemetry. The GCS can also send instructions to the drone. A GCS has many forms.
- **LOS:** Line of sight. CAA and FAA regulations require your drone to stay within a pilot's direct visual control.
- **LiPo:** Lithium polymer battery.

AUTOPILOT

I am going to start with the autopilot, as this is the brains of the entire project. There are many different autopilots on the market, but the one I have chosen for this drone build is a Pixhawk. You will find that many basic autopilots are similar in makeup; what sets the Pixhawk apart is its ability to operate with a GCS called Mission Planner. Mission Planner is open source software that you will learn about later.

The flight controller keeps your multi-rotor stable while you are flying in manual mode, and can take over in autonomous mode allowing you to fly to 3D waypoints. What you will find helpful with the Pixhawk is its layout. All the various components fit into the clearly marked ports on the board, making assembly fairly straightforward.

Author's Note: I am led to believe that the name *Arduino* comes from a bar in Ivrea, Italy, which was frequented by the students that created the open source computer hardware and software way back in 2005. ArduPilot was just one of the many Arduino-compatible boards that were designed for autonomous navigation of model aircraft, boats, and ground-based vehicles. ArduCopter is an open source platform created by DIY Drones community to facilitate the flying of an unmanned aerial vehicle. The original board was the ArduPilot, which came out in 2009, swiftly followed by the APM 1/2, 2.5, 2.6 in subsequent years 2010 through 2012. The Pixhawk arrived in 2012, an incredible achievement in just a few short years.

PIXHAWK SPECIFICATIONS

Processor:
32-bit ARM Cortex M4 core with FPU
168 MHz/256 KB RAM/2 MB Flash
32-bit fail-safe co-processor

Advanced sensor profile:
3 axis 16-bit ST Micro L3GD20H gyro used
for determining orientation
3 axis 14-bit accelerometer and compass
for determining outside influences and
compass heading
MEAS MS5611 barometric pressure
sensor for determining altitude
Built-in voltage and current sensing for
battery condition determination
Connection for an externally mountable
UBLOX LEA GPS for determining
absolute position

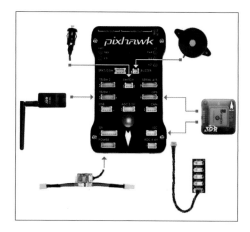

Pixhawk and its main components. Note:
Should you find the Pixhawk a bit too much for
your budget, you could use an APM 2.6
instead—as the instructions vary very little.

Power:
Ideal diode controller with automatic failover
Servo rail high-power (7 V) and high-current ready
All peripheral outputs over-current protected, all inputs ESC protected

Interfaces:
5x UART serial ports, 1 high-power capable, 2x with HW flow control
Spektrum DSM/DSM2/DSM-X Satellite input
Futaba S.BUS input (output not yet implemented)
PPM sum signal
RSSI (PWM or voltage) input
I2C, SPI, 2x CAN, USB
3.3 and 6.6 ADC inputs

Dimensions:
Weight: 1.3 oz. (38 g)
Width: 2 in. (50 mm)
Height: 0.6 in. (15.5 mm)
Length: 3.2 in. (81.5 mm)

ANCILLARIES

The Pixhawk comes with a lot of ancillaries, and it is wise to use them all. The kit contains:

- Buzzer
- Safety switch button
- 3DR power module with XT60 connectors and 6-position connector cable
- Extra 6-position cable to connect a 3DR GPS+Compass module
- Micro USB cable
- SD card and adapter
- Mounting foam
- 3-wire servo cable
- I2C splitter module with cable

The front of the Pixhawk, showing the connection points for your components. The mini SD card is on the forward edge, with the USB connection on the side. The ECS and radio connection pins are all on the bottom edge by the Pixhawk name.

These will be explained later in the assembly. However, to complete your component list, I recommend that you purchase an SD card for your Pixhawk. Most of these ancillaries are simple to understand. For example, the arm button is an illuminated switch that you press as a final safety feature to arm the drone. The fact that it is illuminated also provides a visual status of the drone's safety. Be aware that this is only *part* of the arming process, more of which will be covered later in this book.

The buzzer is a small audio alarm that issues a series of sounds to let you know what the autopilot is doing. It is important to learn at least some of these sounds, as they will indicate if your drone is ready to fly or if there is a problem.

The SD card is essential to record the flight data of the drone. Data is recorded on the drone and can be played back later using a function in Mission Planner. The SD card is particularly useful if you have had a crash and to help discover what went wrong. Under normal circumstances logging starts with arming the vehicle, but only active flights are interesting for analysis. Again, there will be more on this later in the book.

SENSORS

At this stage it is important to understand what the Pixhawk does. It is fundamentally the brain of your drone; in reality it is a PCB board with a central processing unit, sensors, and connectors. The sensors are the most important part of the autopilot since they perform operations that keep the drone level and stable in the air. The connectors are there so you can attach vital inputs to the autopilot such as RC receiver, GPS, compass, and telemetry.

The sensors consist of the following: accelerometer, gyroscope, Inertia Measurement Unit (IMU), compass, magnetometer, and barometer, which all work to keep the drone stable. In addition to these components, we should also consider the GPS to be a sensor as this accounts for our drone's whereabouts on the earth's surface.

Although you don't really need to know what all the autopilot sensors do, if you're a novice it's good to learn a brief description and have some basic understanding.

Accelerometer

Accelerometers measure linear acceleration in up to three axes (let's call them X, Y, and Z). A very important characteristic of three-axis accelerometers is that they detect gravity, and as such can know which direction is "down," as gravity is a fixed constant.

Gyroscope

A gyroscope measures the rate of angular velocity change in up to three angular axes (let's call them alpha, beta, and gamma). Together with an accelerometer, they help keep your drone stable in flight.

Inertia Measurement Unit (IMU)

An IMU is essentially a small board that contains both an accelerometer and gyroscope (normally these are multi-axis). Most contain a three-axis accelerometer and a three-axis gyroscope, and others may contain additional sensors such as a three-axis magnet.

Compass / Magnetometer

An electronic magnetic compass is able to measure the earth's magnetic field and use it to determine the drone's compass direction (with respect to magnetic north). This sensor is almost always present if the system has GPS input and is available in some GPS units.

Pressure / Barometer

Since atmospheric pressure changes the farther away you are from sea level, a pressure sensor can be used to give you a pretty accurate reading of the UAV's height. Most flight controllers take input from both the pressure sensor and GPS altitude to calculate a more accurate height above sea level. Note that it is preferable to have the barometer covered with a piece of foam to diminish the effects of wind over the chip.

OTHER MAJOR COMPONENTS

In addition to our autopilot, we will need several other components in order to get our drone flying nicely in the air and under full control. For example, the autopilot needs an airframe to sit on, and motors and propellers to make it fly; we also need to communicate with our drone as well as know its position.

FRAME

The frame is the physical body on which all our components will be placed. We will be using a simple X configuration known to most as a quad-copter. As already mentioned, a drone can have three, four, six, or eight arms; it all depends on your design. Without a doubt the quad-copter (four arms) is the most common configuration, although there are many others. For example, we can add four more motors under the existing ones (one top and one bottom) and make it an octo-copter. Having more than the basic four motors does come in handy should you have a single motor failure, preventing the drone from simply falling out of the sky. Six- and eight-motor copters will, for the most part land safely if there is a failure.

However, I find the basic quad-copter is harmonious while allowing the simplest principle of operation for controlling roll, pitch, yaw, and motion. This combination provides superb hover capabilities, quick maneuverability, as well as speed. For example, if the same components were placed on a + configuration frame, you would only have one motor pushing in each direction, whereas with an X configuration we have two. We will use a DJI F450, which is mentioned in the build process.

MOTORS, PROPELLERS, AND ESCs

For the purpose of this book I am going to wrap our motors, propellers, and speed controllers into a single entity called the propulsion system. There has been a lot written and discussed in the model world regarding which motor, propeller, and speed controller combination to use. For a "newbie," it's a subject to avoid for your first build, as I have spent years trying various combinations to achieve the best thrust-to-weight ratio. At the end of the day, the aim is to use a combination that will lift your drone into the air and keep it there as long as possible. It also helps if you have enough thrust to fight any strong wind. For our drone build we will use a pre-defined propulsion system, the DJI E300, which I have used on several models before with great success. I have found that this kit provides excellent reliability while giving outstanding thrust-to-weight ratio. The system is great for drones with a takeoff weight between 3¾ and 5½ lbs. (1½ and 2½ kilos), including the battery.

> **Author's Note:** As an approximate guide the recommended takeoff weight when using a 3S LiPo battery is ¾ lbs. (350 g) per motor and 1 lb. (400 g) when using a 4S battery. As we are using a 3S battery with four motors, our rough takeoff weight will be between 3 and 3½ lbs. (1.4 and 1.6 kilos).

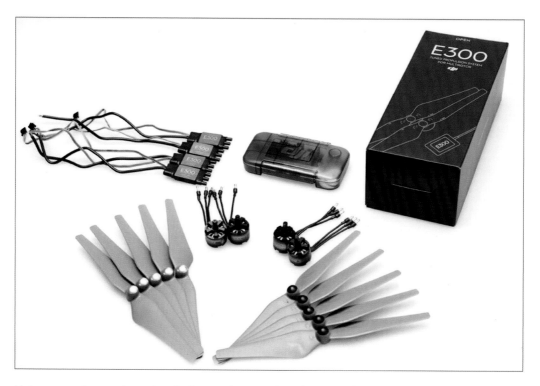

Motors, propellers, and speed controllers are best purchased as a match combination based on the type and final weight of your drone.

What comes included with this E300 kit?
4 x motors
4 x propellers
4 x ESCs
1 x accessory pack/toolbox. This includes the screws for mounting your motors to the frame.

The propulsion system has four DJI 2212 920kv motors which are of excellent quality for the price, and, once again, I have found them to be extremely reliable and efficient. They are designed to take self-tightening propellers, a mechanism that dramatically reduces any chance of loosening rotor or mistakenly installed rotors. Additionally, tool-free mounting and dismounting makes storage and transportation easy.

DJI 9443 Self-Tightening Propellers

These propellers have been designed to be as aerodynamic as possible. They are finely balanced through expert manufacturing and reduce the "jello" effect when a camera is fitted to the drone. Cutting-edge design focused on dynamic balancing means the flight controller is able to achieve a higher gain value to produce better stability combined with precision agility.

DJI E300 ESC 15a (Mark 11)

These ESCs are designed to be used as part of your propulsion system and have integrated sensors that are capable of transmitting real-time diagnostics. Brand new sine-wave technology reduces power consumption and will increase the efficiency of your quad-copter agility. These powerful ESCs can manage to maintain a steady course in the most challenging of environments. The E300 can be used with either a 3S or 4S LiPo battery.

GPS

A GPS unit in the drone will provide us with its location. Additionally we can use the same GPS unit to send the drone to a location, that is, a longitude and latitude. Storing locations, such as where you actually took off from, means you can use this to request the drone to return, should there be a problem such as a broken communications link, using Return to Launch (RTL). Adding GPS to a drone has seriously increased the usability and operational capacity, allowing the user to perform some pretty amazing feats.

For our build we will be using the new 3DR u-blox GPS with built-in compass, which has now been updated to a u-blox NEO-7 GPS module. I have chosen this GPS module because it is entirely compatible with the Pixhawk autopilot. It also provides precision accuracy coupled with a quick warm-up time. The unit from 3DR comes complete with two cables, one for the GPS and one for the compass.

Battery

Drone battery power has increased dramatically over the years with new chemistry being launched that provides even longer flying time. This is what battery weight, size, and output power are all about. The object is to match your overall drone weight (including the battery) with the best possible power source. Ultimately, we want to take off and fly with as much thrust and duration as physically possible. For our drone we will be using a 3S 11.1v 2200mAh battery with a 30c discharge for our initial build; later on in this book we will switch to a larger battery when we design our own 3D-printed body.

Lithium polymer batteries are the preferred power sources for most drone hobbyists today, as they offer high discharge rates and a high energy storage/ weight ratio. Lithium polymer batteries, commonly known as LiPos, have been with us for many years, but their care and maintenance are paramount as they can prove extremely dangerous. Added to which they are very expensive, and a LiPo battery that is well treated will last longer. If you purchase a drone with a battery, most likely it will come with a matched charger. There is a lot written about LiPo battery care, but the basic rules are as follows:

- Only use a charger that is specifically designed for LiPo batteries.
- Make sure your charger is set to the correct voltage and cell count.
- Use a good quality charger that balances each cell within the pack.
- Always charge your batteries in an open ventilated place.
- Do not leave your battery charging unattended (many people do). They can catch fire.
- Place the charger and battery on a surface that will not burn in the event of a fire.
- If the battery starts to swell and bloat, switch off the power and disconnect. Do not use this battery and dispose of it safely.
- Check your battery to make sure it is not damaged or punctured, especially after a crash.

Transmitter and Receiver

The transmitter and receiver are used to send and receive control messages and other vital information between the operator and the drone while in flight. For example, if I want my drone to fly in a particular direction, then I simply push the appropriate stick on the transmitter, and the receiver gets the message and passes it to the autopilot action—hence the drone moves to my commands. Transmitters and receivers come with a number of channels, with each channel being assigned to a particular function. For our transmitter and receiver, I am going to suggest two different models. One is fairly cheap and the other is more expensive, but the latter could save you a lot of money should you decide to fit a camera to your drone.

First Choice: Orange RX

The great thing about this transmitter is that it is cheap and just packed with features you would only find on a more expensive unit. The Orange Rx T-SIX is a superb, fully programmable 6-channel 2.4 GHz DSM2-compatible transmitter that is loaded with features and ideal for flying our quad-copter. This little gem even has an adjustable RF power output setting in the menu, allowing users to switch between US and EU power standards. All the programming functions are displayed on a large and easy-to-read backlit LCD display. You can easily navigate through the menus with the convenient scroll wheel or by using the menu buttons. The menu system is very intuitive and easy to use, unlike some other transmitters on the market today that require studying the manual to operate.

It uses four standard AA batteries but also has a built-in JST plug so you can also run it off a 2-cell LiPo battery. The transmitter's low voltage battery alarm is adjustable via the menu to suit whichever battery chemistry you choose.

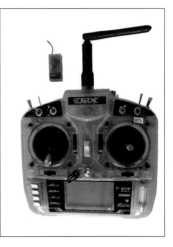

Orange RX T-SIX and receiver, cheap but very good value.

Features:

- It has a 3-pos Flap/Gyro switch, which is ideal for use with multi-rotor aircraft as a flight mode selector switching between Stabilize to AltHold and Loiter
- Compatible with all DSM2 receivers
- 6-channel operation
- Supports 3 wing/tail mixes (dual aileron/elevon/V-tail)
- Supports 2 swashplate types (1 servo 90°/CCPM 120°)
- 5-point graphic pitch/throttle curves
- Adjustable gyro gain
- Dual rates and exponential
- Servo reversing (all 6 channels)
- Channel mixing
- Sub trim and travel adjust
- Flaps and differential
- Throttle hold
- Large backlit LCD display
- 10 model memory
- Adjustable RF power output (US and EU select via menu)
- Integrated timer
- Adjustable stick length
- Trainer port

Specifications:

Frequency: 2.4 GHz
Modulation: DSM2

No. of channels: 6
Model memory: 10
Stick mode: Mode 2 (left stick throttle)

Orange RX R620 2.4 GHz DSM2 Compatible Receiver

If you intend to use the Orange RX transmitter, then I would personally recommend the R620 receiver. It is small in size and weighs ⅓ oz. (10.2 g). This 6-channel receiver offers full range and is easy to bind and operate.

Walkera DEVO F12E is a RC transmitter and video receiver all in one. It even allows you to change the angle of the camera while in flight.

Second Choice: Walkera DEVO F12E

Although this is my second choice, it is only that way because of cost. I own two DEVO F12Es, and they are simply brilliant. That said, you get a lot of transmitter for your money. Why is it so expensive? Well, it is a long-range first person view (FPV) transmitter with a host of functions, including camera control. This is the most advanced transmitter made by Walkera, which has years of experience in the hobby industry. The DEVO F12E uses a 2.4 GHz control system with 5.8 G image transmission function. Though it's a little on the large side, the weight and feel of the transmitter is excellent. The grip rubber and stick controls are really well placed, as are the camera gimbal controls.

While the 5-inch LCD color screen is excellent, you will need the included sunshade when working in bright sunshine. This gives you fantastic live video range with LOS out to around half a mile (800 meters). There is also a telemetry function providing a basic on-screen display indicating range, GPS data, battery usage, and lots more. While the specification provided indicates an RC range of just under a mile (1½ kilometers), I have regularly achieved well over 1¼ miles (2 kilometers) with no issue.

Specifications:
Encoder: 12-channel micro computer system
Remote control frequency: 2.4 GHz DSSS
Image frequency: 5.8 GHz
Image channel: 32
Output power: ≤100mW
Current drain: Close the video ≤300mA (100mW) / View the video ≤430mA (100mW)
Power supply: Li-Po 7.4V / 11.1V 1600–3000mAh or 5# battery 8*1.5V or NiMH
    8*1.2V 1600–3000mAh
Output pulse: 1000–2000 US (1500 US neutral)
Remote control distance: 500–1500 m depending on LOS and terrain
Working time: 4–6 hours
Weight: 780 g

Devo RX703A Compatible Receiver

For reasons that will become obvious later in the build, we are going to use a RX703A 2.4 GHz receiver in our drone. This is a 10-channel small unit weighing in at just around ⅓ oz. (11 g) It is simple to use with a DEVO F12E transmitter and provides excellent signal strength.

Although soldering is probably the one skill at which you will need to become proficient, there are also some other refinements we should look at before assembly. Balance is just one; we need to keep the weight evenly distributed from the center. Moreover, as we are planning to use several radio frequencies, we need to be careful about where we place each component.

Telemetry

Telemetry refers to a communications system between the GCS and the drone much in the same mode as the RC transmitter and receiver mentioned earlier. Depending on the frequencies used, telemetry can have a slight edge over the RC range; additionally, it offers what is known as *automatic flying*. This simply means that via a software GCS, it is possible to plan a mission, press a button, and the drone will carry out your instructions, from automatic takeoff to automatic landing. The mission might include making way-points at which the drone could carry out a secondary function such as shooting video or still images. It might also include circling an object with the camera always pointing at that object. A full range of automatics features can be found later in this book.

SKILLS NEEDED

Knowing what a component does is not all you need to know. You also need to know how to solder, and how to prevent too much vibration or frequency interference. This means we have to consider these factors during our build. We must make sure our solder joints are strong with no chance of a short in any circuit. It is equally as important to annul as much vibration as possible because vibration affects the delicate sensors in the autopilot. Likewise, the wrong positioning of various communications may cause frequency interference.

SOLDERING IRON

During our drone build we will be dabbling into the basic skills of soldering and the world of electronics. Soldering might seem a simple process, but it takes time and skill to become proficient.

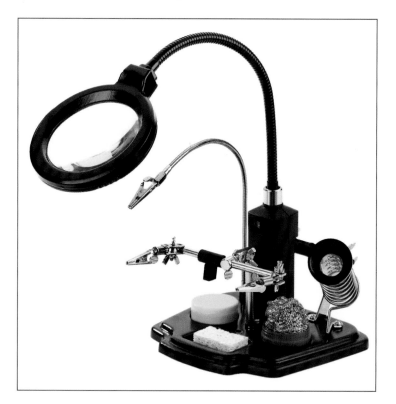

You can used a simple soldering iron to do the work, but it is far better to use a proper setup that allows you to see (using lamp and magnification) your work while it is being held using additional arms.

A soldering iron is a tool with a metal tip that heats up hot enough to melt metal solder. It is used to transfer heat to things like wires or electronic components being fitted to a printed circuit board (PCB). After the appropriate areas are heated properly, solder is applied. Soldering irons vary in the amount of heat they produce, and it is critical to get the right type for the work you are doing. For example, too much heat applied to a PCB can sometimes damage sensitive electronics already installed.

Since the soldering iron generates a lot of heat, caution is needed when in use. When not in use, the iron should be placed in a cradle. Your cradle can be a simple affair which holds the handle while keeping the heated tip away from contact and out of harm's way. Simply placing a heated iron on its side could result in your burning yourself or setting your workshop on fire. Cradles come in all shapes and complexities; some are plain wire frames while others have controls for adjusting the amount of heat.

Most cradles will have some form of cleaner on which to wipe the iron tip; this is normally a damp sponge, but it can also be a brass sponge. As you use your iron, the tip will oxidize and turn black mainly due to the impurities in the lead-free solder. Wiping it clean on a wet or brass sponge will keep your tip clean and aid efficient soldering.

Solder

Originally, the main component of solder was lead, which is very harmful to humans; luckily, today we have lead-free solder. However, the reason lead was used was its superb ability to act as a joining agent. Lead-free was not quite so adaptable, as the components often require a higher degree of heat. Most lead-free solder contains a "flux core," *flux* being an agent that makes lead-free solder flow easier.

Tips on soldering:
- Handle hot irons with care.
- Wear safety glasses.
- Use medium heat (325–375 degrees C)
- Smoke coming from your iron means the temperature is too high.
- Tin your tip with solder before making each joint.
- Use third-party clamps or vices to aid you solder.
- Use the side of the tip, not the very end.
- Heat the joint or component evenly.
- Remove the solder first and then the iron.

Practice, and then practice some more. If your joint does not hold, look at what you are doing wrong. Start again and keep practicing until you can make a nice smooth solder joint and not a great big ball.

Most of the soldering required in this book involves attaching simple wires to a PCB that has few or no other electronic components that we can accidentally harm. Nonetheless, there may come a time when you want to add other components to a fully loaded PCB, such as when modifying an autopilot. This must be done with extreme care and only when you are completely proficient. While a solder iron and soldering are not too expensive, if you plan to make more drones in the future, you might consider getting a helping hands clamp. This is basically a stand that has a light, a magnifying glass, and a set of clamps that will help grip the parts you want to solder. That leaves your hands free to hold the solder and iron.

Author's Note: Soldering the ESCs and power cables to the bottom board is fairly straightforward, especially if you have a little knowledge of the subject and have soldered before. If not, I would encourage you to at least try. If you feel that it is not within your ability, maybe someone from the local aeromodel club who is good at soldering can help. You should also avoid purchasing an expensive soldering kit if someone else solders for you.

FREQUENCY RADIATION AND ANTI VIBRATION

Just about everything on a quad-copter radiates a frequency; it's not just the radio control or the telemetry, but the motors, autopilot, and GPS are also affected. Then, if we plan to add a camera with an FPV transmitting system, we have to worry not just about the extra frequency but also the vibration. Too much vibration will not just affect the video quality, but can also cause the autopilot sensors to become unstable. There is also some alignment to consider. For example, the compass in the autopilot needs to be pointing in the same direction as the compass in the GPS unit. Paying attention to these points will all help to get your drone in the air and flying well the first time.

Anti-vibration systems will help reduce many of the problems when it comes to flying your drone.

While not wanting to get too technical with our build, I do want to explain the importance of vibration damping for your drone. The Pixhawk autopilot has sensors built into the board that are very sensitive to vibrations. The firmware combines the data from these sensors (accelerometers with barometer and GPS data) to calculate an estimate of its position. The aim of vibration damping is to reduce high- and medium-frequency vibrations while still allowing low frequency actual board movement to take place in concert with the airframe. By damping down the vibrations in the autopilot, we help avoid this problem. The vibrations from your drone can actually be measured, but that is not within the scope of this book.

SUMMARY

As with any mechanical or electrical build, it is always best to know and understand what component part does what. Likewise, every industry has its own related jargon, and it is good to understand terms such as DSSS (Direct Sequence Spread Spectrum) when you come across them. As a newbie to drones, there is no need to go into depth at this stage, but always try to learn what the acronym means. When selecting your RC transmitter, take your time and be advised by others. Your transmitter is an equal part of your drone flying, so become familiar with it and its settings.

Physical skills are also very important, and these will only come with patience plus a lot of trial and error. I have been soldering for years and I still make a mess of it sometimes. Finding someone to teach you how to solder is a real bonus when building your own drone. The instruments used in the construction of an autopilot are extremely sensitive to both vibration and frequency, and it is worth understanding how you can help annul these.

Chapter 4

DRONE ASSEMBLY

Now that we have decided to build our own drone from scratch and have a little understanding of the role each component plays within the construction, it's time to get to work. We are first going to build our drone using a simple X configuration frame, as this allows you to see clearly where all the components go and how they connect to each other. You will notice as the build progresses that your drone starts to look complicated, with various wires crisscrossing the structure. Done worry—it will still fly, and later in this book we will look at producing a purpose-made body and transferring the electronics so that we have a commercial looking product.

Our frame is a DJI F450, but there are many similar frames and clones on the market, such as the Q450 from HobbyKing. The frame you purchase should have four arms of equal length and a top and bottom board with all the various nuts and bolts. The bottom board of the DJI F450 also comes ready for ESC wiring, and if you purchase a different frame you should check that it, too, has similar features. While it is possible to buy a four-arm quad-copter body of different form, for this build we will stay with the standard X configuration.

Author's Note: You will find several different ways of assembling a drone in other books and on the Internet. Some people like to add the basic components to the autopilot and connect to the software in order to upload the firmware first and so forth. I would discourage you from this practice, as the procedure of calibration is much smoother if all the parts are fitted securely to the frame first.

CONSTRUCTION

Our first task is to solder the ESCs onto the base board and then assemble the frame and add the motors. After this, we will position our autopilot on an anti-vibration system before adding all the peripheral components such as GPS, telemetry, and buzzer.

Your frame kit should consist of:

4 x arms (normally two red and two white)
1 x bottom board (with integrated power circuit)
1 x top board
2 x packs of screws (M3x8 16 - M2.5x6 24
1 x Powerline pair (red and black cable)

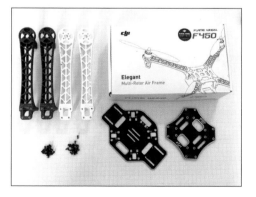

Layout of a F450 frame.

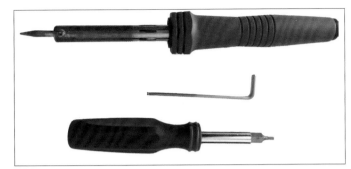

Tools required for assembly.

For assembly you will require the following tools:
2.0 mm hex wrench
Soldering iron and solder

As previously stated, during a build we find lots of extra unwanted wires, so before we solder our ESC wires to the base board it is best to try and gauge the required length of ESC cable you are going to need. Most motors and ESCs come with a lot of extra cable that only gets in the way during the build. So the first thing to do is lay out the bottom board and attach any one of the arms. Next, connect the three wires from one of the motors to any one of the four ESCs. Place the motor in its position at the end of the arm and stretch out the ESC cable to see how much spare cable you have when it reaches the soldering point on the bottom board. Remember that you will need to thread your cable through the arm later, so do not cut it too short. Best to do a physical check first to determine how much to cut off. Once you have gauged the required length, remove the attached arm so that the bottom board can lay flat, ready for soldiering.

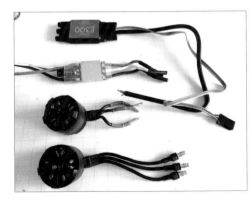

Cut cables to the required length, but always allow a little extra as it's easy to shorten a cable but impossible to lengthen after cutting. Every ounce you save will increase your drone's performance.

It is also a good idea to solder a battery connector to the end of the two power cables before attaching it to the board. There are several types of connectors, but we will be using a fairly standard XT60. If you later intend to add an FPV camera system, you might also consider at this stage adding the power supply connector, as this will save a lot of time and trouble later.

Author's Note: One thing that has really annoyed me over the years is the range of different battery connectors. While others will argue over which is best, I find it beneficial, especially if you have several types of drone or fixed-wing aircraft, to stick to one particular battery connector—for me this is the XT60.

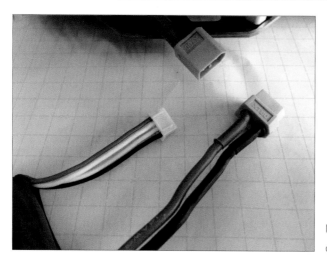

Male XT60 on the drone to female battery connector.

Make sure your soldering iron tip is not too large or small, and allow it to become hot enough to melt the solder wire. Touch the center of each solder pad on the bottom board with the solder wire and apply the soldering iron so that you have a neat small lump of solder on each pad.

Apply a small amount of solder to each end of the ESC power cables. Next, simply touch each ESC cable in turn to its relevant pad on the bottom board [red wires on (+) and black wires on (-)] and apply heat with the soldering iron. Once all wires are firmly soldered to the board, I like to cover them with tape or better still a little hot-melt adhesive to lower the risk of a short. Take your time and make the soldering as neat as possible while ensuring a good, solid joint.

Next you should decide which way is "front" on your quad-copter; to do this, we are going to place the two white arms forward and the two red arms at the rear. Although this is not really necessary, you will find it a tremendous help when it comes to flying your drone and visualizing orientation.

Next, assemble the frame by screwing the arms onto the top board first, followed by screwing down the bottom board. I find it easier to flip the frame over on its back to screw in the bottom board, which also helps me thread the ESC wires through the arms. To stop the ESCs from moving about, attach them to the arms with a small plastic cable tie. Your frame is now ready to accept the motors.

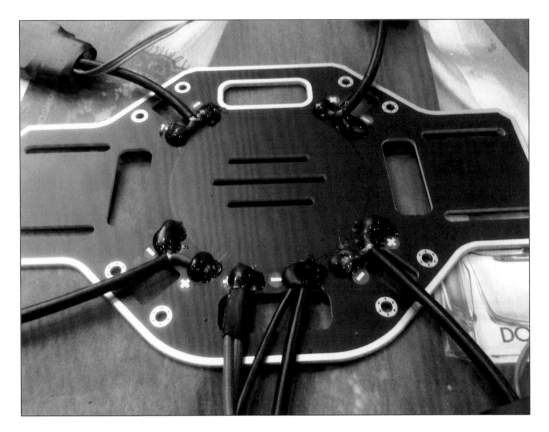

Soldering the ESCs and covering the joints for protection against a short.

Before we fit our motors to the frame, we need to check the rotation. The DJI 310 Tuned Propulsion System comes with two CW (clockwise) and two CCW (counter-clockwise) motors, which have a different direction spin-on thread. The correct propeller will only fit on the correct motor. You need only do this for one motor and one propeller. Depending on which motor you choose (CW or CCW), place that motor on the relaxant arm. Remove the propellers and screw your motors into place using the four screws in the following order. Make sure your screws are tight—but do not over-tighten:

CCW motor on the front right arm. Number 1.
CW motor on the front left arm. Number 3.
CCW motor on the rear left arm. Number 2
CW motor on the rear right arm. Number 4.

Assembled frame with ESCs and power cable.

Thread the motor wires down through the legs and connect the bullet pins to the ESC. Try to keep the same configuration on the motor ESC connections for all four arms. Next, connect the bullet connectors of the ESCs to those of the motors. But make sure there is enough slack in the cable so you can unplug the bullet connectors between the motors and ESCs in order to correct any rotation direction (explained later). Additionally, as the ESC cables move under the framework, it is best to number them on the autopilot connector ends. This will help when you come to connect them later on.

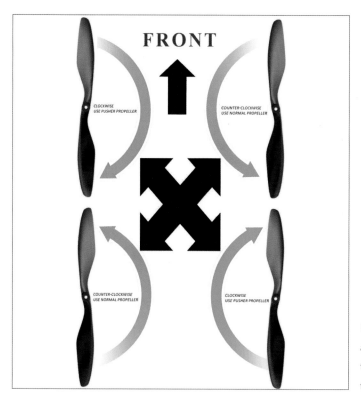

Motor rotation is important and I always remember direction as the front motors always turn in towards each other.

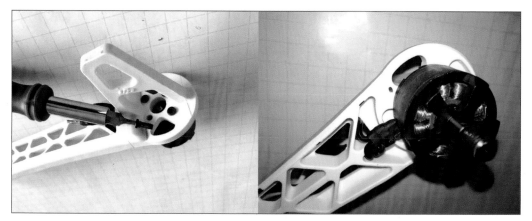

Use the screws provided to fit your motors and thread the wires through the arms so that they are ready to connect to the ESCs.

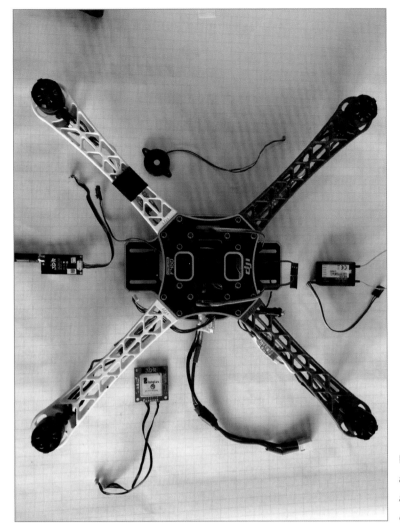

Frame, motors, and ESCs assembled. Start thinking about the additional components' positioning.

Your basic quad-copter frame should now look something like the picture above— if so, you are now ready to receive the autopilot and the other components. Remember what I mentioned in Chapter 3: anti-vibration, balance, and position- ing of external components are important.

Author's Note: While there has been a massive improvement in autopilots and components, there seems to have been little attention paid to the actual frames. Given that it is wise to reduce electrical interference by distance, in the past I have added a small flat board to my frame at this stage. I use a small flat carbon sheet which I bolt under the top frame. This makes no dif- ferent to the flight but allows room for my components and helps reduce interference.

The Autopilot

Your autopilot is probably the most expensive part of your drone build; while I could have chosen a cheaper option, believe me, it's worth the money. The Pixhawk is an amazing autopilot for one very logical reason: it is extremely advanced in that it will perform autonomous flying including such functions as Follow Me, and it is incredibly simple to use. Additionally, it is a mature platform that runs on excellent open source software. This software allows us to calibrate, tune, and actually fly our drone. So let's get started!

Anti-vibration is extremely important and will reduce many of your flying problems. This applies not just to your autopilot but also to your cameras when mounted.

The first thing we need to do is position the autopilot on the frame. For the purpose of this build, we are going to mount the autopilot on the top board and hold it in position with anti-vibration tabs. I like to use 15 x 15 x 10 mm soft silicone gel tabs, as I find these really work well. You can also use a purpose-made anti-vibration board.

First place your frame on a flat, level surface with the two white arms (front) pointing away from you. Then attach one side of the anti-vibration tabs to each corner of the autopilot. You will notice that the top case of the autopilot has a large, white arrow; align the autopilot with the top board of our frame with the arrow pointing directly forward and press down firmly. Your autopilot should be placed as close to the center of gravity as possible, as this will provide the best sensation of movement and eliminate some flight problems later.

Author's Note: In the future, you may need to, orientate your autopilot in a different position. Mission Planner offers over thirty possibilities. This will involve changing the Roll, Pitch, and Yaw values in the Advanced Parameters screen. However, for our build we will leave it pointing forward.

Next we will plug in the 3DR power module that should have come with your Pixhawk. Simply plug the 6-pin cable into the Pixhawk where it says Power. This will supply a steady 5v supply and allow the unit to measure the current and voltage of the battery. The one end of the power cable can now be connected to the XT60 on the bottom board. This will now leave you with a XT60 male connector ready for the battery. (Do not connect the battery at this stage.)

Connect the XT60 power module that comes with your Pixhawk pack.

That's it; you are now ready to start attaching the ESCs, radio control, and peripherals to your autopilot. Before we start, we need to consider where we will place our components. For example, I like to position the Safety switch where I can press it without the fear of the motors suddenly starting. I also like to keep my radio receiver and telemetry unit well apart to avoid any electrical interference. When placing your attachment, keep in mind later additions such as a video transmitter or LIDAR unit.

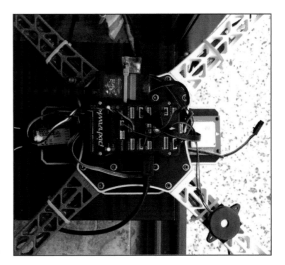

All components placed, plugged into the autopilot, and secured.

Connecting the ESCs

Next we will connect the ESC wires to the autopilot. Recall the advice on labeling each connector, as it can get a bit confusing with the tangle of wires. The procedure is extremely simple.

Connecting the ESCs to the autopilot.

Motor 1 plugs into number 1 on the autopilot; the 3-pin connector is Negative / Positive / Signal top to bottom. (I mention this because some ESCs use different colored wires.) Repeat for motors 2, 3, and 4. These simply clip onto the autopilot at the end with all the pins in the section marked MAIN OUT. You will see on the side that the rows of pins are marked - + S; these correspond to the three wires of the ESC: - brown is ground -, orange is Power +, and yellow is signal S. There is a small groove at the bottom of the autopilot case so that you can only feed the ESC connector in one way.

1. Thread the ESC cable from Motor 1 (front right) to the number 1 on Main Out.
2. Thread the ESC cable from Motor 2 (rear left) to the number 2 on Main Out.
3. Thread the ESC cable from Motor 3 (front left) to the number 3 on Main Out.
4. Thread the ESC cable from Motor 4(rear right) to the number 4 on Main Out.

> **Author's Note:** Now you will begin to notice the amount of cables and wires suddenly hanging around. Although you should not worry too much about this, it is a good idea to keep your wiring as neat as possible. A second point to remember is that all electronics give off a frequency that, in some cases, can cause interference, which can affect performance.

As a guide, I have placed the various peripherals on the frame as shown in the figures. While this configuration is not carved in stone, it has worked well for me in the past.

GPS

The only real exception to placing your components on the top board or attaching them to the arms is the GPS. Your GPS *must* see sky, and allow the compass to be distanced from interfering magnetic fields. It is possible to place your GPS very close to your autopilot, provided you use some form of shielding, but we will come on to that later. For the moment, we will mount our GPS on a small pole where it is clear of all the electrical power running around the frame. I recommend using the 3DR UBlox GPS with compass, as this works well with the Pixhawk and has been tested by thousands of users worldwide.

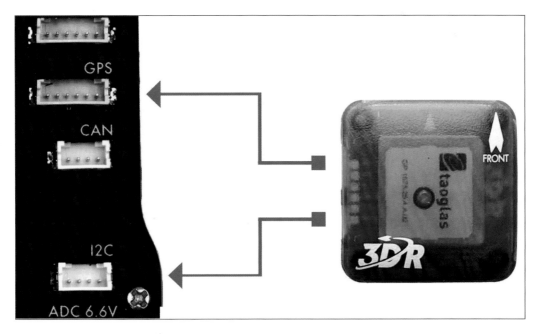

Connect the 6-pin connector from the GPS to the port on the Pixhawk marked GPS. Then do the same for the 4-pin compass connector which fits into the I2C port.

> **Author's Note:** If you intend to use other devices that will connect to the I2C port later on, such as LIDAR, you will need to use a I2C splitter. This is simply a small board that extends the number of devices that can be attached to the I2C port.

Use your GPS mount to elevate its position as far away as possible from the motors and ESCs. Make sure your GPS is at least 4 inches (10 cm) away from the battery, its wiring, or any other metallic objects. Use only plastic or aluminum to secure the GPS mount.

Now we will lay out the various components on the frame, keeping in mind that the cable supplied must easily reach its port on top of the autopilot. The connectors are very small, so please take care when fitting them.

I2C splitter.

Radio Receiver

I have used double-sided Velcro to stick my RC receiver onto my extra board just to the right of the autopilot. The two aerial tails I like to fix with small ties to the front arms. The radio antenna should radiate a good downward signal to ensure good connectivity over distance. Plug the cable into the RC pin slot on the autopilot.

Telemetry Receiver

Telemetry is basically two radios that communicate instructions and send information between them. Depending on which country you live in, the legal frequency range will vary. There are two common frequencies for this communication: 433 MHz (mainly Europe) and 900 MHz (mainly North America); but there are other frequencies such as 868 MHz (Europe) that can make a huge

The radio receiver is plugged into the side pins to the left of the ESC connections marked RC. The telemetry unit clips into the connector marked TELEM 1.

difference to range. You should check which is permitted in your country. The telemetry radios can be calibrated to work together using the utility within Mission Planner. Once complete, one unit is installed into the drone and the other is connected to Mission Planner or DroidPlanner.

I always like to place my telemetry receiver unit on the opposite side from the RC receiver, again using Velcro to attach it to my extra board. As with your RC receiver, your telemetry antenna should be well placed to receive the best signal. Plug the cable into the autopilot marked Telemetry 1.

SAFETY BUTTON

As luck or design would have it, the Safety button will actually fit nicely into one of the slots on the top cover. Screw it on and make sure it's secure. Plug the cable into the autopilot marked Switch. The Safety switch is an additional safety feature of the Pixhawk. Normally drones are armed by pushing the throttle stick down and to the left or some similar combination. When powered, the Safety switch will flash red (the full function is described later in this book).

BUZZER

The buzzer is best situated away from the autopilot because, when it makes a noise, it creates a lot of electrical interference. Best practice is to connect the cable to the autopilot into the slot marked Buzzer and then use the maximum range of the cable and attach it to an arm using a tie.

Safety button and buzzer attached.

Author's Note: You will have noticed by now that most of the components that fit into the autopilot have small connectors that simply push in one way. Should you need to remove one of the components for any reason, be extremely careful. Pulling the connector out of the autopilot takes some skill, and it is easy to rip the wires out of the connector.

With all our peripherals securely in place, we are now ready to add the battery. For our build I have first chosen a smaller 2200mAh 11.1v 3s battery, because it is lightweight. Later on, when the drone is flying well, I will increase this to a 3350mAh 11.12v 3s battery. Both batteries will slide neatly between the top and bottom boards of the frame and will be held in place by a Velcro strap. When we connect the battery, you will notice that a lot of lights start flashing on your drone. These flashing lights tell you a lot about the drone's condition. The main LED is the large one just below the arrow indicating front. This will flash in a series of colors and combinations explained in the following figure.

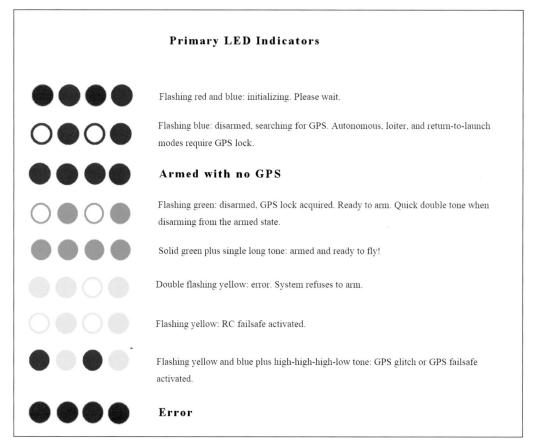

Primary LED Indicators

Flashing red and blue: initializing. Please wait.

Flashing blue: disarmed, searching for GPS. Autonomous, loiter, and return-to-launch modes require GPS lock.

Armed with no GPS

Flashing green: disarmed, GPS lock acquired. Ready to arm. Quick double tone when disarming from the armed state.

Solid green plus single long tone: armed and ready to fly!

Double flashing yellow: error. System refuses to arm.

Flashing yellow: RC failsafe activated.

Flashing yellow and blue plus high-high-high-low tone: GPS glitch or GPS failsafe activated.

Error

Battery connected light combinations.

You will also notice that lights will appear on your RC receiver and telemetry receiver, and a range of different sounds can be heard—which is all good.

BINDING THE TRANSMITTER AND THE RECEIVER

One of the first things we want to do is to make sure that the transmitter we are using is talking to the receiver on the drone. Since we have built our drone so that the receiver is exposed, we can do this while it is attached to our drone.

> **Safety Note:** If you have placed the propellers on the motors, you should remove them for safety.

Switch on the transmitter and setup a new model for your drone, that is, give it a specific name. You can read the instructions on this in the manual but the next section gives some basic instructions.

ORANGE RX SYSTEM USING R800X DSMX COMPATIBLE 8-CHANNEL RECEIVER

Make sure your transmitter is off, all switches are in the back position, and the throttle is fully down. You will need a *bind* plug to be placed in the receiver. In the case of the Orange receiver, this is clearly marked "bind." Simply slip the plus onto the pins.

Step 1. Connect the battery to the drone; the light on the receiver should start flashing.
Step 2. With the transmitter off, flip the "training" switch forward and hold in position.
Step 3. Switch on the transmitter; the screen will light and show the word "BIND."
Step 4. Wait until the light on the receiver has stopped flashing and gone solid.
Step 5. Disconnect the drone battery and remove the bind plus.
Step 6. Release the "training" switch and power off the transmitter.
Step 7. Switch the transmitter back on and reconnect the battery to the drone, and the receiver light should come back on and stay solid.

WALKERA DEVO F12E

As with the Orange Rx system, we will need to set up a model name prior to binding the receiver to the transmitter. Make sure your transmitter is off, all switches are in the back position, and the throttle is fully down.
Step 1. Connect the battery to the drone, and the light on the receiver should flash continuously.
Step 2. Power on the transmitter, and the receiver light should go solid.
Step 3. Disconnect the battery from the drone.
Step 4. Power down the transmitter.
Step 5. Switch the transmitter back on and reconnect the battery to the drone, and the receiver light should come back on and stay solid.

It is not unusual for the transmitter and receiver not to bind on the first attempt, and you may have to repeat the process several times before having success.

With the transmitter and receiver working, it's time to boot up with the firmware and software and configure the drone. For this, we will need to connect the drone to the open source software called Mission Planner.

ESC Calibration

Once your build is complete and you are ready to add the battery and power the drone up, you may find that not all the motors are spinning—this is mainly due to the ESCs not being calibrated. Almost all ESCs need to calibrate to read the minimum and maximum pulse width modulation signal. You can do this by calibrating all four motors at the same time or individually. ESC calibration varies based on what brand of ESC you are using, and you are advised to read the accompanying documentation. As we have changed our ESCs to Castle, the procedure is fairly straightforward:

- Check that your RC transmitter is bound to the receiver.
- **Remove** the propellers. (Do not do this with the propellers attached.)
- **Remove** any USB cable if connect to PC or laptop.
- Turn on the RC transmitter and push the throttle up to maximum.
- Attach a **full** battery to the drone. (The LED on the autopilot will flash red, blue, and yellow indicating that the ESC calibration will begin the next time you connect your battery.)
- Leave the throttle stick at maximum **high** and disconnect and reconnect the battery.
- Press and hold the Safety button until it displays solid red.
- The autopilot is now in ESC calibration mode, and the LED will flash red and blue.
- You will hear a number of beeps depending on the battery you are using (three for 3S and four for 4S); you will also hear two more beeps indicating the maximum throttle has been captured.
- Pull the throttle down completely to minimum, and you should hear one long beep indicating the ESC calibration is complete.
- Test the motors by raising the throttle a little up and down.

SUMMARY

This is the most important chapter in *Build a Drone*, and it's vital to get these steps right. Building a drone is a straightforward matter of construction with plates, arms, nuts and bolts, screws, and Velcro ties.

If you can handle a knife and fork and avoid cutting yourself shaving, then you have the skill to build a drone. Check that you have all your components ready and practice a little soldering prior to working on your drone. Remember to "tin" the solder spots on the frame and the tips of the wires so that when you join them all that's required is heat. Make sure there are no solder joints touching. If you make a mistake when connecting the components to the autopilot's slots, be extremely careful when removing the cable because the wires can easily come out of the connector.

Weigh everything. When building small to medium drones, it is key to keep the weight down to an absolute minimum—and here I am talking ounces. If you can find a component that will do the same job at a much lighter weight, you simply have to give it a try. Saving the odd ounce here and there results in better and longer flying. In a build of a medium drone weighing around 1 pound (500 grams), saving 2 ounces (50 grams) will make a vast improvement.

Never take the manufacturer's specification weight as gospel. When you place the actual components on the scale, they may be 4 or 5 grams heavier, and keep in mind that there are cables to be calculated as well. When I first produced a drone similar to the one shown in this book, I managed to remove over 3 ounces (85 grams) of surplus cable.

At this stage of the build, you are best advised to keep the propellers *off* the drone except for when you want to test motor direction (battery removed). Also, get into the habit of removing the battery from the drone before turning off the RC transmitter.

The real challenges lie ahead: learning about your autopilot calibration and settings and learning to fly the damn thing, but this all adds to the fun.

Chapter 5

SETUP AND CALIBRATION

To calibrate and fly our drone we need a software application that runs on a laptop computer (Mac or PC), desktop Mac or PC, tablet, or smartphone and communicates with the drone via a USB cable or wireless telemetry. There are several different applications for ground control stations (GCS), all of which should work with your Pixhawk-based drone. These include Mission Planner, DroidPlanner 2, APM Planner 2, and Tower (DroidPlanner 3). Now that our drone is ready to be calibrated, we are going to be using the open source GCS software Mission Planner.

If there is one piece of open source software that has helped push the evolution of both commercial and hobbyist drones in the past five years, it is Mission Planner. Developed by Michael Oborne (with contributions from many other extremely clever people), Mission Planner is a Windows-compatible program that is used to configure a drone (it will configure just about any drone, helicopter, fixed-wing aircraft, or ground based robot). Not only will it upload the firmware and software that controls your drone, but it will also provide an autonomous mode so that your drone will take off and land automatically as well as fly through way-points. Mission Planner can help tune your drone so that it flies at optimum performance, or allow you to analyze your flights and correct the drone when things go wrong. It provides Google Maps (plus many other maps) and the ability to use first person view (FPV). Simply saying that Mission Planner is impressive does not do it justice, and you should consider making a donation if you intend to use it. Some of the outstanding features of Mission Planner include its ability to do the following:

- Load the firmware (the software) into the autopilot that controls your vehicle.
- Set up, configure, and tune your drone for optimum performance.
- Plan, save, and load automatic missions into your autopilot with simple point-and-click way-point entry on Google or other maps.
- Download and analyze mission logs created by your autopilot.
- Set fail-safe procedures in the event of unexpected problems.

With appropriate telemetry and video hardware you can:

- Monitor your vehicle's position and status while in operation.
- View live video downlink on-screen.
- View the heads-up display (HUD).
- View and analyze the telemetry logs.
- Operate your vehicle in FPV.

The download page, together with license and safety requirements, can be found at firmware.ardupilot.org/downloads.

You will need to download and install the Mission Planner software to either your laptop or desktop before we can proceed.

> **Author's Note:** I own a Windows computer and therefore have listed the instructions as such, but the same program is available for those using Macs.

Mission Planner has been developed over a number of years and at first glance might look a little complicated. However, its wealth of functions is outstanding, and with a little practice you can learn so much about drone flying and fault finding, which all helps to build a safer flying knowledge base. This book does not allow for a full explanation of all the possibilities within Mission Planner, but I have highlighted those that are most important.

The first screen to appear when Mission Planner loads is the Flight Data screen. This screen is divided into several sections. At the top you will see a navigation bar with page controls. At the top left are icons for Flight Data / Flight Plan / Initial Set-up / Config / Tuning / Simulation / Terminal / Help and Donate. At the top right of the screen are two drop-down boxes and a Connect button. Initially we are interested in the first three screens, and the rest we will cover later. This navigation bar remains on all screens so you can navigate between pages. Right-clicking on the strip will allow you to auto-hide it.

> **Author's Note:** We will only be using the first four screens since the aim is to get our drone into the air and see how it is performing. However, Mission Planner offers a complete solution with fine tuning, and you are encouraged to explore it more once you have learned the basics.

THE FLIGHT DATA SCREEN

The Flight Data screen is used for primary flying, showing location, and drone attitude. The HUD is in the top left corner. The HUD displays your artificial horizon, heading direction, bank angle, and ground or air speed, among other things. Two of the most important pieces of information shown in the HUD are your GPS status and your current flight mode.

- Air speed (ground speed if no airspeed sensor is fitted)
- Crosstrack error and turn rate (T)
- Heading direction
- Bank angle
- Wireless telemetry connection (% bad packets)
- GPS time
- Altitude (blue bar is rate of climb)
- Air speed
- Ground speed

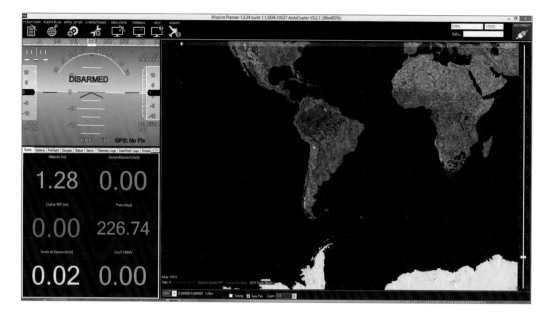

Mission Planner Flight Data screen.

- Battery status
- Artificial horizon

You can add more items to your HUD by right-clicking the screen and selecting the required user items from the list. You can also use your HUD window as an FPV screen when the appropriate equipment is connected. You can do this in the Configuration window and select any attached video input from the drone down list. Then click start.

The main window is the Mapping window, which normally loads Google satellite maps by default. You cannot see your drone on the map until you have connected and have GPS lock, at which stage the map will move onto your location. There is a slider on the right to zoom into the map. If you right-click on the Mapping screen, you will see a list of commands, most of which are flight commands. The Mapping screen has a lot of other features that can be seen by pressing Ctrl+F, but I will cover these later.

The third window is at the bottom left and provides a number of different displays such as Gauges, Status, Telemetry Logs, and more. It will open in Quick mode, which indicates altitude, ground speed, yaw, vertical speed, and so on.

THE FLIGHT PLAN SCREEN

This is where you can plan your flights and script other actions you want your drone to carry out. Note that the flight plan screen is mainly all mapping with the top navigation bar and a way-points bar at the bottom. To the right of the

Mission Planner Flight Plan screen.

screen is the Action window; this shows the mouse cursor GPS location in various formats. You can select from a large range of mapping types, load, save, read, or write a way-point file and set the home location.

Right-click the Mapping screen to show the list of functions. This list helps build a set of way-points or instruction functions that can be saved or uploaded to the Mission Planner program or drone. One of the more important functions is the Pre-Fetch, which is found under Mapping Tools. This allows you to cache maps in Mission Planner when you have Internet available and use them outdoors when Internet is unavailable. You can select your preferred mapping format from the drop-down menu and set your home location.

THE INITIAL SETUP SCREEN

The Initial Setup screen is where we start. It allows us to install the latest firmware, calibrate our drone, and set up other hardware such as our telemetry. Having installed Mission Planner and taken the time to look over its features, we will remain on the Initial Setup page. It is a good idea to make sure your computer speakers are turned on and the volume is high enough to hear the verbal instructions.

On the Initial Setup screen you will see Wizard in the left column. Click this to bring up the Mission Planner Setup Wizard screen.

Select your vehicle, in our case a Multirotor, and click Next.

Highlight your frame type, X (top left), and click Next.

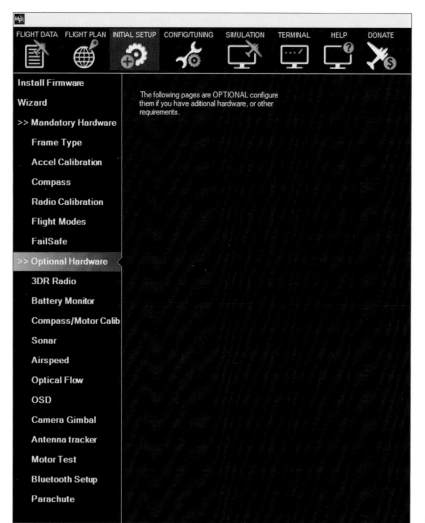

Mission Planner "Initial Setup" screen.

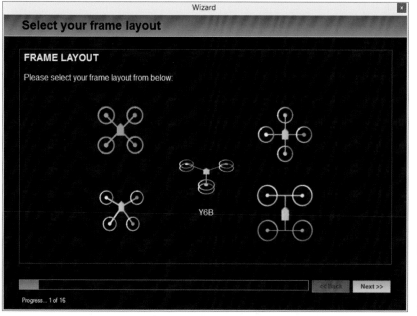

Select the Setup Wizard and follow the on-screen menu.

You will now be asked to connect your drone to the computer via a USB cable. This will enable you to update the firmware on your autopilot and carry out the full calibration. I would suggest you use a long USB cable, as you will be required to move the drone around in various directions as part of the calibration process. Be careful not to disconnect the USB cable until the whole Wizard process is complete.

Author's Note: All the lights should start flashing, and it is a good idea to check that each and every peripheral is getting power; normally this is done with a red light, most of which may be blinking. Ideally, position yourself so that you can see the computer screen and your drone is close to the window where it can eventually receive a good GPS location. I say "eventually" because it can take some time for the GPS receiver to get sufficient signals to maintain a 3D GPS lock. Remember that a GPS that is new, or one that has changed location over a long distance, will take some time to orientate itself.

At this stage you'll see a comms port appear, and the firmware update will start. When finished, you will see a separate screen asking you to wait until the sounds have stopped. Mavlink will now connect your drone to the software. Click Next.

You will be asked to select your frame layout—in our case it's the X frame. Highlight and click Next.

The following screen requires you to calibrate your accelerometer. Click Start and follow the on-screen instructions. This involves positioning your drone flat, on its side and nose, etc.

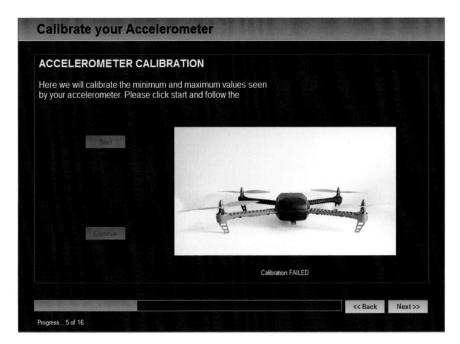

Accelerometer Calibration

During accelerometer calibration you will be asked to place your drone in various potions such as lay it on its side, nose down, and so on. There are an on-screen illustrated menu and verbal instructions to follow.

Calibrate your compass. Click Live Calibration and then rotate the drone in every axis, that is, twist it round and round, over and over, and circle it in every direction. Do this until the screen says you're finished—you'll hear a single beep, and you'll get a small screen with New Mag Offsets. Click OK and then Next. Beware that you may have to do this for several minutes; take care not to disconnect the USB cable during this calibration. You will be informed when the calibration is complete.

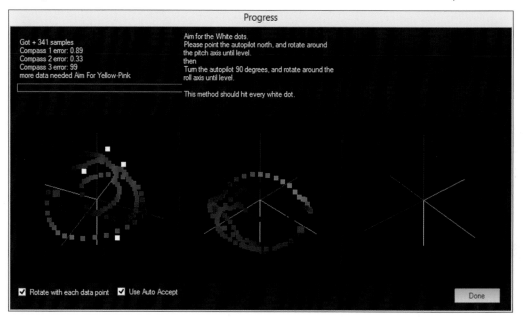

Compass Calibration.

The next screen is the Battery Monitor screen, and you will be asked which auto-pilot version and sensors you are using, plus what battery you intend fitting. We are using the Pixhawk and a 3DR Power Module with a 3500mAh battery. Click Next.

At the moment we have no Optional Items, so click Next.

RADIO CALIBRATION

If you did the bind procedure correctly at the end of the last chapter, this next step should be easy. Your drone is already connected, and the light on the RC receiver should be flashing slowly. Switch on the RC transmitter, and the light on the receiver should go solid. Click Continue.

Author's Note: Your drone can be controlled using only an RC transmitter or through a combination of RC and a laptop or tablet ground station. Always calibrate your RC using the Mission Planner configuration utility. The GCS joysticks are set in Mode 2, which means the left joystick is throttle. When taking off and landing manually, the drone is normally in Stabilized mode.

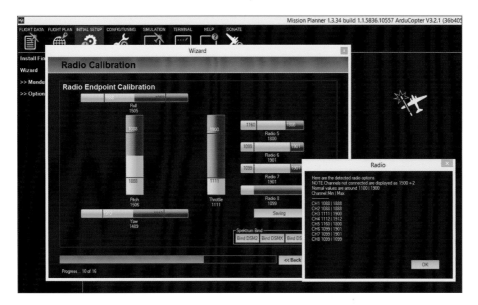

Radio calibration complete.

The Radio Calibration screen in Mission Planner will transform from all grey bars to grey and light green bars, with readings by each bar. Click the Calibrate Radio button, and you will receive a "warning" to remove the propellers and make sure your battery is *not* connected. Click OK, and you will be asked to move the

RC flight modes.

joysticks to their limit in all directions (I generally push them up to each corner). You will see the green markers move as you do this. Check that the throttle moves the throttle bar, the elevator the elevator bar in the corresponding direction, and so on. Also check that your Mode switch is working by flicking it back and forth on the transmitter while watching the screen.

It will ask you to click when done. Make sure all your sticks are centered and the throttle is in the down position and click OK.

RC FLIGHT MODES

The final thing we need to do is calibrate the flight modes on our RC transmitter. On the Flight Mode page, you will see a list of modes, one of which is normally highlighted. If you flick the Mode switch on your RC transmitter, you will see the highlighted window move up and down the list; for example, it might start in Stabilize mode and move through AltHold and RTL. If you click the down arrow on the right side, you will see the drop-down list from which you select your flight modes. I like to set them as Stabilize / AltHold / Loiter. Once you are happy, click Save and move on to the next page.

Flight Modes

The next window will show flight modes, which are controlled by the RC transmitter switches. You will see Flight Mode 1 is Stabilized, and this is highlighted. If you click the small arrow box to the right, it will bring down a list of flight modes available. *Do not change* from Stabilized until you have learned a lot more about flying.

For the moment, I recommend you leave the flight modes in Stabilized, AltHold, and Loiter as this will ensure you do not accidentally set your drone in a mode from which you cannot easily recover. You will also see there are boxes for Simple and Super Simple modes, which are described later in this book.

Flight modes are a key feature of drone flying, and you will find them when using your RC, a laptop, or tablet to control your drone. A detailed description of your flight modes comes later in this book, but briefly the most important ones are as follows:

Stabilized: This holds the drone in a stabilized position where you can use the RC controller to adjust flight with the trim switches. If your drone flies well in Stabilized mode, you will have very little trouble. You do not need a GPS lock to fly in Stabilized.

AltHold: Altitude mode uses the barometric pressure to hold the drone at a given height. The drone will still drift a little and respond to any strong wind gusts. You do not need a GPS lock to fly in AltHold.

Loiter: Loiter uses GPS to lock the position of the drone. You will need your GPS unit to acquire at least six satellites (eight or more is better) before you engage Loiter mode. The more satellites, the better the lock and drone stability.

Auto: Choosing Auto means putting your drone into automatic mode. You need GPS lock with at least six satellites (eight or more is better) before you engage Auto mode. When you engage Auto mode, the drone will search for any flight plan and attempt to carry out the mission.

> **Author's Note:** To this day I am extremely cautious of Auto mode, as it can lead to a lot of problems and the possibility of a flyaway or crash. In Auto mode, the drone is acting on a set of instructions; if those instructions are incorrect or you have not assessed the area over which your drone will fly, you could be putting your drone and others at risk.

RTL: Return to Launch is also a major mode, as this will bring your drone back to where you took off from. This is very useful if you lose orientation or see your drone flying into trouble such as towards a building or tree.

Guided: When you are at your flying field or flight location and have a good GPS 3D lock and a strong telemetry link, take off in Stabilized mode, and once you have reached your desired height, switch to Loiter mode. On the Flight Data screen, right-click and select Fly to Here; you will be requested to enter an altitude in meters. A Guided mode flag should appear at the point requested, and the drone will fly to this location and await further instructions. You can now switch to another mode or instruct it to fly a mission in Auto mode.

> **Author's Note:** Guided mode can only be used when you have a GCS (Mission Planner) with a Mapping screen and telemetry, as you cannot use this mode when flying with just an RC transmitter. (Some of the latest RC transmitters with a FPV screen have got on-screen commands.)

Acro Mode

This is not a mode for beginners, and you will need a lot of pilot training before you venture into Acrobatic mode. Basically, you use the joysticks to fly, and it is designed for aerial acrobatics such as flips and drone racing. Because we are using Pixhawk, an Acro Trainer parameter can be set either on/off to make learning to fly in this mode easier.

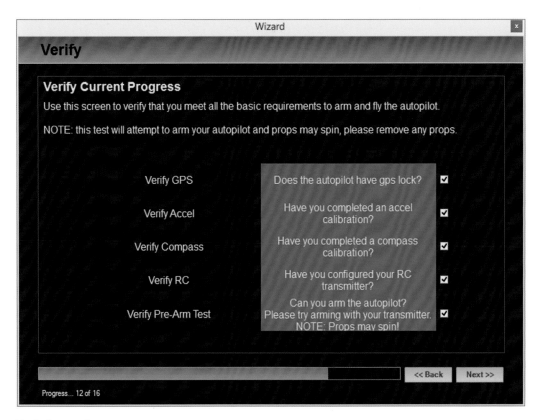

Verify Current Progress

This is set in COnfig/Tuning—APM: Copter Pids using Channel 7/8 switches choosing Acro Trainer. Leaving the 3-position switch in the off position will disable the Trainer, while the middle position will enable automatic leveling, and the third position limits leveling and lean angle. More advanced tuning can be found at ardupilot.org/copter/docs/acro-mode.html.

The next page of the Wizard is to verify that the setup process is complete and that you are left with a drone capable of flying. This includes the Pre-Arm Test in particular, which, for a variety of reasons, might prevent you from arming your drone prior to flying. We are looking to get green across the board. If there is a failure, it will need to be addressed before you can arm the drone. A list of pre-arm failures will be checked and discussed before we do our first test flight.

FAILSAFE

The Failsafe window in the Wizard allows you to set certain fail-safe actions that your drone will automatically take when there are problems. No matter how well you build and calibrate a drone, there are always things that can go wrong. A fail-safe is a function that is activated by default when something unexpected happens, such as loss of GPS or a break in the communications between the RC GCS and the drone. Using Mission Planner, it is possible to provide instruction to the autopilot that in the event of a problem a set fail-safe will automatically activate. When you first look at the Failsafe settings in Mission Planner, they can look a little confusing, but you are encouraged to look at all the options under Config/Tuning, Standard, and Advanced Parameters as they soon become clear.

LOSS OF GPS SIGNAL

In all my years of using GPS, I have only experienced a few instances where the signal has not been available. A loss of signal is generally caused by poor pilot skills or trying to fly in a place where the GPS signal is weak or blocked—indoors, under thick forest canopy, and in extremely poor overcast weather conditions are good examples. Make sure the GPS on your drone has a good GPS lock by moving to an open area where it can see lots of sky; do not fly when the sky is extremely overcast with mist and fog; avoid flying under power pylons as this can temporarily kill your signal.

If your drone loses GPS lock for more than 5 seconds in a mode that requires GPS (Auto, Loiter, RTL, Circle, Way-point navigation), it will initiate a Land or AltHold, depending on your settings. In Mission Planner, GPS Failsafe is enabled to Land by default, and I highly recommend you to leave it enabled.

Failsafe Options

LOSS OF COMMUNICATION

A more likely scenario is a loss of link between your RC transmitter and the drone. If you have set your fail-safe correctly, your drone will return to the launch point (RTL) before landing. You also have the option to instruct the drone to hold its position (AltHold) or land (Land)

Depending on what you are doing at the time of a communications failure will, your fail-safe settings will activate different modes. After a loss of communication of more than 5 seconds while flying in manual mode, the drone will land if no GPS is available; if GPS is available, then the drone will RTL; if you are flying in Auto mode and you have GPS, the drone will continue on its mission and hopefully regain communication as it gets closer to the RC.

LOW BATTERY

Low battery is another fail-safe you can plan for. As your battery reaches 25 percent of charge (or whatever it is set at), the drone will automatically land or return to the launch point depending on your fail-safe instructions.

GEOFENCE

When you start flying, it is always best to confine your flights to an open area and at a distance where you can observe the drone and its altitude. A *GeoFence* is

GeoFence

an area on the map that is fenced off and prevents your drone from going outside its boundaries. We use Mission Planner to instruct the behavior of our drone when the fence is reached. For example, if the drone strays outside the set fence boundary, it will switch to RTL or land; likewise if it flies too high, the GeoFence will activate. The minimum recommended fence radius is 100 feet (30 m), but you should set your height and boundaries well within the legal flying limits issued by the local aviation authority. Your drone must have a good GPS signal in order to operate correctly, so do not disable the GPS during your pre-arming checks. Finally, make sure you set the RTL sufficiently high enough to avoid any obstacles should the GeoFence activate.

> **Author's Note:** While the GeoFence is a great idea—especially when learning to fly—I must freely admit that I have never tried it.

At the end of the Wizard, you will get a window indicating where you can get extra information, including some excellent advice on measuring vibration and fail-safe mechanics.

OPTIONAL HARDWARE

Once you have finished with the Wizard, you can check out the optional hardware. This allows you to check on your battery, sonar, or an airspeed indicator as well as items such as antenna tracker or the release of a parachute. It is worthwhile looking at each in turn.

At the moment, the only optional hardware we are using is the telemetry, so we need to click the 3DR Radio. With the drone connected via a USB cable to the computer, select the comm port and click Connect. Then select 3DR Radio in the Initial Setup window, and if the two telemetry radios are compatible, you should see the reading appear in the screen. Most telemetry radios come in pairs, so you should have no problem.

CONFIG/TUNING WINDOW

This is an important window and a major part of the Mission Planner software, because it allows you to customize the PID (Proportional, Interval, Derivative) and various critical settings that make your drone fly better. Here you can adjust your flight modes, confirm or change your GeoFence, and do some basic or extended tuning. Most importantly, you can enable and disable your hardware should the need arise. The window also allows you to adjust both basic and advanced parameters with a list explaining what each of the parameters handles. At this stage you should not attempt to adjust any parameters until you fully understand their functions and the effects they have on the drone's flight and stability.

Simulation

The Simulation screen is available in Advanced view. This screen is used in conjunction with Xplanes and Flightgear (fixed-wing only) and jMAVSim (multicopters only) for Hardware in the Loop (HIL) simulation when used with your board. Although I am sure this is an excellent simulation program, I find it extremely difficult to set up and use; for beginners, there are much simpler simulation programs you can use.

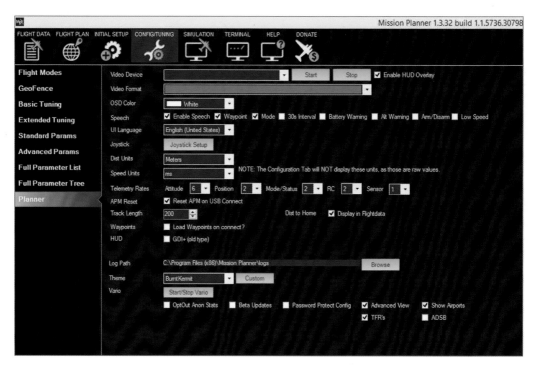

Config/Tuning Window

Terminal

In the Terminal screen, you can manually set up your APM and run tests on sensors, log readings, and other functions. You will need to select the autopilot currently attached to Mission Planner and click Connect. You will now see a list of scripts showing the current status of the autopilot. If there are any logs available for download, then you can view these and analyze them.

Author's Note: As you are a newbie to drones, my advice is to leave the Terminal screen untouched; however, it is a great aid to discovering what happens if your drone is not flying as it should or the reason for a crash.

Finally, disconnect the drone and remove the USB cable.

UNDERSTANDING YOUR RC CONTROLLER

Building and configuring your drone is only half of what you need to learn; flying safely is equally as important, and you cannot fly a drone without a controller. For our first test flight, we will be using an RC transmitter. RC transmitters come in a wide variety of shapes and sizes, but most have something in common: two joysticks. I have mentioned that we are using either an Orange Rx T-SIX or a Walkera Devo F12E for our project, and so I will explain the latter because it has the most features.

Before you do any flying, you really need to get to know your RC transmitter and how it works. Despite your eagerness to get your drone up in the air, you first should read the RC transmitter manual. The RC transmitter is a handheld controller that lets you pilot your quad-copter and control its movements. When you make an adjustment with the joysticks, it sends a signal to your quad-copter instructing it on what movements it should make.

THE LEFT STICK CONTROLS THE THROTTLE AND YAW

Throttle is increased by pushing the left stick forward and decreased by pulling the left stick backward; in doing so you change the altitude/height of the quad-copter. All four motors increase or decrease in speed, providing positive and negative lift. When flying, your throttle will be engaged constantly until you land—it is important to make sure it is adjusted correctly.

Yaw is adjusted by pushing the left stick to the left or to the right. This simply rotates the quad-copter left or right, allowing you to point the quad-copter in the desired/changed direction. Depending on direction, two diagonally opposite motors increase or decrease in speed causing rotation.

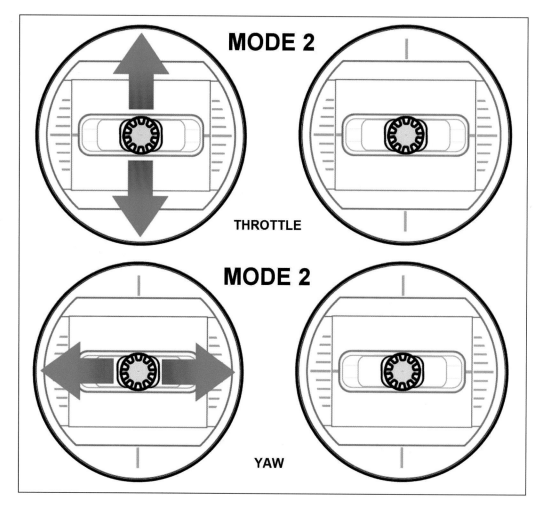

Throttle and Yaw

THE RIGHT STICK CONTROLS ROLL AND PITCH

Roll is adjusted by pushing the right stick to the left or right. The quad-copter rolls left or right in the direction of the joystick movement—with a quad-copter, this is more of a sliding action. Depending on direction, the two motors on one side increase or decrease in speed causing the quad-copter to move in the direction of the decrease (least resistance).

Pitch is created by pushing the right stick forward or backward. This tilts the quad-copter and makes if fly forward or backward. Depending on direction, the rear or front motors increase or decrease in speed, causing forward movement in the direction of the decrease (least resistance).

When you start flying, there is a tendency to push the sticks too hard in all directions—especially when you feel you are losing control of the quad-copter. Remember that the harder you push the stick, the faster your quad-copter will move. Get in the habit of using gentle, small movements until you have become proficient.

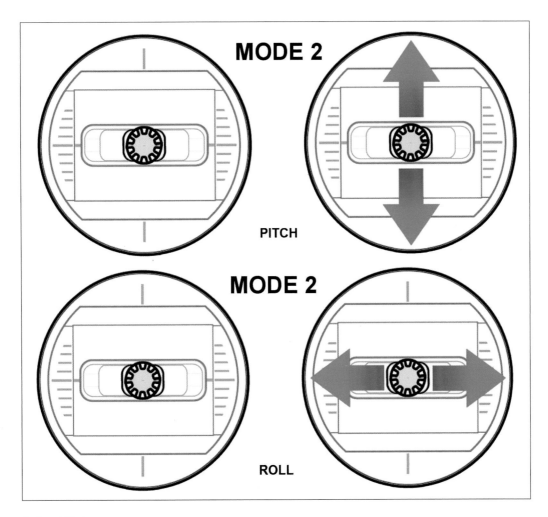

Roll and Pitch

TRIM BUTTONS

When you first handle your RC transmitter, you should check that all the trim settings are centered; most RC transmitters emit a *beep* sound when you center the trim. Each of the joystick movements has its own adjustment, called trim. When first test flying your drone, the moment you leave the ground you may find that it starts to drift in one or more directions. To counter this, we need to move the trim switch in the opposite direction of the drift. For example, if your drone is drifting forwards you need to reduce the pitch trim.

Using the trim control to correct stability.

Intelligent Orientation Control (IOC)

You may come across the term IOC, which is a mode that helps when learning to fly. One of the great difficulties with quad-copters or any multi-rotor drone is orientation, especially when the drone becomes so small that it is difficult to see which way it is flying. If the drone has reversed so that it is pointing toward you, and you try pulling back on the pitch stick, you will be sending the drone further away.

IOC is designed to counter this by using the pilot's location as the fixed point so that all stick movements are relevant to that position. In short, no matter the heading of the drone, if you pull back on the right hand joystick it will come back towards you. With the Pixhawk autopilot, the same function is offered using Super Simple mode.

Super Simple mode allows you to control the drone relative to its direction from where it took off (armed) and therefore relies on a good GPS lock. This mode can be assigned to Channel 7/8 switches on your RC transmitter.

FIRST TEST FLIGHT

By now, we should have our drone calibrated and ready for a test flight. When we carried out the full Wizard, we also did a pre-flight check. Providing it all went well, and you are confident that your RC transmitter and receiver are working, we should have no problems moving ahead. But as we have fitted a telemetry unit, we will continue using Mission Planner to aid us with our first flight.

Choose an open area away from structures and habitation for your first flight; best of all, go with someone who is a qualified pilot. Your local model airfield is a great place to start, as they will have chosen a safe place to fly and have lots of experienced pilots available.

Author's Note: This is my local model airfield in Spain (it is forbidden to fly anywhere else), which supports both a full concrete runway and a helicopter/quad-copter takeoff pad. It has full power facilities and a lot of open country around it with a large reservoir to the east. If you have to travel any distance to your flying field, make sure both your drone and RC transmitter are safe and secure. You should always take your propellers off when in transit and make sure they will not be damaged.

Local airfields are the best places to test and start your flying career.

FIRST FUNCTION TEST

As this is a new build, we're going to do a basic function test. This test is simply the same procedure as if you were going to actually fly—but with the propellers removed.

Check the drone and make sure it's all okay with everything secure, because we don't want components falling off in flight.

- First switch on your RC transmitter, making sure it has a sufficient power supply.
- Double-check that the throttle is down to zero and all switches are in the back position.
- Place a fully changed battery into the gap between the top and bottom plates of the frame and secure with Velcro strap. Make sure it's as central as possible.
- Place your drone so that it is sitting on level ground.
- Connect your battery, listen to the sound indicating all is well, and note the LED lights sequence. The motors may twitch a little.

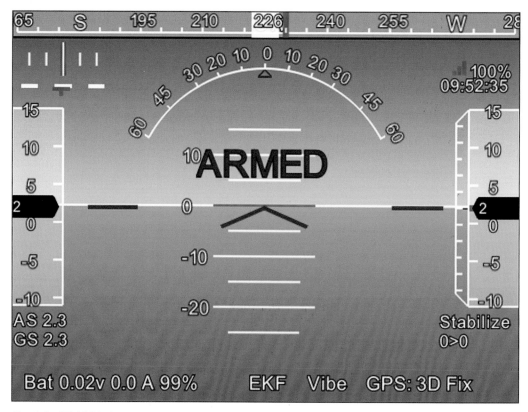

Check for 3D GPS lock—you can see this on both the autopilot LED and Mission Planner/DroidPlanner if it's connected.

Author's Note: At this stage it is a good idea to look at what the drone is doing, such as checking what color is the main LED on the autopilot. Is the drone making any unusual noises? Is there a flashing or static light on all the components: radio receiver, telemetry, ESCs, and Safety button? If you have Mission Planner on a laptop at the field, check to make sure you have a good telemetry link. This will give you confirmation that the pre-arm is good to go and that you have a 3D GPS lock.

If all seems okay, press the flashing Safety button until it turns solid (about 2 seconds is normal).

Next, arm your drone using the RC transmitter by pushing the throttle down and to the right. You should hear a sound indicating the drone is armed.

Gently push the throttle up slowly, and the motors should spin up; reduce throttle and then increase a second time to make sure they are all working. Note if the direction of rotation is correct.

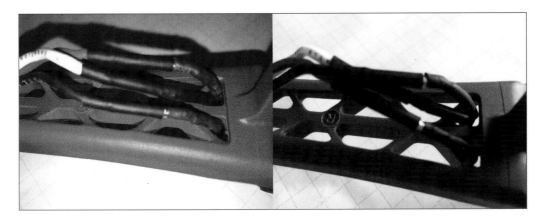

Check motor rotation; if it's wrong, change any two wires between the ESC and the motor.

Fully lower the throttle and push and hold to the left to disarm. Press the Safety switch on the drone until it is flashing. Then disconnect the battery. Finally, switch off the RC transmitter.

Place the propellers on the drone and carry out the same procedure as before. This time, when the drone is armed, apply sufficient throttle to *almost* lift the drone, and test all the joysticks to see if the drone is responding with the correct movement. Slowly push the throttle up to provide sufficient thrust so that the drone almost takes off (wobbles a little). Then push the left-hand stick (throttle) to both left and right (yaw) a little and see if the drone responds by leaning slightly in the same direction as your joystick movement. Do the same for the right-hand joystick (roll and pitch) and make sure the drone is responding as it should. Once you are satisfied, increase the throttle steadily but in a positive manner so that the drone breaks contact with the ground and flies into the air about 6½ feet (2 m) in height. Hold at this position.

Correct and trim as instructed before; try to keep the drone balanced as you do this. If the drift is strong, then I suggest you land the drone and adjust the trim and take off again. Once the drone is flying stably, reduce the throttle slowly so that the drone descends and lands; hold the throttle down and to the left to disarm the drone. When the motors have stopped, approach the drone and press the Safety button until the red light starts flashing. At this stage you can practice several more short takeoffs and landings before moving onto the flying lessons in Chapter 6.

Author's Note: While your drone build should work correctly if you have followed the instructions in Chapters 4 and 5, there is the possibility that one or more pre-arm checks prevent you from actually flying. If this is the case, consult the pre-arm faults so you can resolve them.

We used Mission Planner to set up our drone. Now we are going to look at how we can use it to resolve any issues and also assess what went wrong should the drone suddenly crash or behave improperly.

Having spent a good deal of money on the drone, the last thing you want is to have a "flyaway" the first time you connect the battery. To help prevent this, Pixhawk uses a range of pre-arm safety checks. These checks cover a wide variety of things such a bad sensors and missed calibration, plus a whole host of other issues including compass alignment. While these checks are important and help prevent crashes and flyaways, most can be disabled if necessary (though that is not advised at this early stage).

Once your drone is assembled and calibrated, you would expect to be able to fly it. When we come to the pre-arm checks in Wizard and all is green, then you should have no problem. However, experience has shown me that you might come across several pre-arm checks that prevent flight. You will discover something is wrong when it comes to arming your drone and there is no response and the LED is flashing yellow. Let's take a look at pre-arm checks and how we can resolve any issues.

You will need to connect your drone to Mission Planner using a USB cable or telemetry. For this I use a large 10-foot (3-meter) USB cable that allows me to place the drone in a position where it can get a good GPS signal. Press the Connect button in the upper right corner and then switch on your RC transmitter before pulling the left-hand stick fully down and to the right in an attempt to arm the drone.

> **Author's Note:** There are quite a few pre-arm faults that can appear, but for the purpose of this book, these are the main ones that I have come across.

Provided nothing has gone wrong, your drone should now be ready to take to the sky.

COMMON PRE-ARM FAILURE MESSAGES

If there is a failure in the pre-arm, it will show in the HUD. Be aware there may be more than one fault. You must resolve any faults one at a time until your drone will successfully arm.

First off, your drone will not arm until you have a 3D satellite fix, so your drone needs to be in a position where it can get at least six or more satellites.

RC NOT CALIBRATED

Solution: Make sure your RC transmitter and receiver are calibrated as described in previous chapters.

COMPASS NOT HEALTHY

Solution: Because this is a sign of hardware failure, try another GPS unit first; if the problem is not resolved, you may need to change the Pixhawk autopilot.

COMPASS NOT CALIBRATED

Solution: Recalibrate your compass.

Compasses Inconsistent

Solution: This simply means the internal and external compass alignment is off; make sure the external compass (arrow on the GPS) is facing the same direction as the arrow on the Pixhawk.

GPS HDOP

Solution: This means there is a problem with the measurement of position accuracy. Without getting too complicated, simply move your drone to a better location to receive GPS.

Acells Not Calibrated

Solution: This is due to poor calibration, so just recalibrate your drone.

Once you have resolved all your pre-arm faults, reconnect to Mission Planner and try arming your drone once more.

SUMMARY

By now you should have a good understanding of how to configure and adjust your quad-copter using Mission Planner. This software is an invaluable asset in so many different ways, and I encourage you to examine all the aspects and functions available. However, I would caution you not to fly in Automatic mode until you have mastered basic manual flying using your RC transmitter. No matter what type of transmitter you use, always read the manual and familiarize yourself with the joystick movements and switches that perform the mode switching from Stabilize, AltHold, Loiter, and RTL in particular.

Seek advice before you test fly your drone and, if possible, have a trained quad-copter pilot on site to assist you. Never take your new build out into the back garden and hope that it flies perfectly first time; use a dedicated airfield or a place away from people and obstructions. Check the weather conditions for your first flight and do not attempt to fly when there is strong wind. Choose a place for your first flight where there is grass as opposed to a hard surface, which will help save your drone should you accidentally land it hard. When you take off for the first time, always use Stabilize mode and try to hover so that the drone remains in the same place. Rock the drone first to confirm that the joysticks are in sync with the drone's movements. Fly safely and progress with the flying lessons.

Chapter 6

SAFETY AND FLYING YOUR DRONE

While there is still quite a lot to learn, now that you have built your drone you will want to fly it. As with our build process I suggest we take the flying part slowly, working our way from basic flying to becoming proficient. Luckily, a quadcopter is just about the easiest type of aircraft to fly, but this can produce over-confidence, making the possibility of a crash highly likely.

I know that many readers may be tempted to skip over the safety briefing, but it could help save your drone (and a lot of money should it hit someone accidentally); additionally, flying it without care can also get you in serious trouble with the authorities. No matter what happens, when we take to the air with all our fail-safes in place and with the drone totally operable, what happens if our drone does something we did not predict? We are forced to deal with the consequences even if we did not intend for the incident to happen.

While hobbyists have been flying safely for years, they usually confine themselves to a dedicated model airfield where strict rules have always been in place. But the arrival of VTOL aircrafts have changed the way in which we look at flying. True, model helicopters have been around for some time and have fitted in nicely with the model aircraft society. It's the new breed of drones that have flooded the market which seem to pose a major problem. Literally, anyone can fly a quadcopter, and while the packaging might say "for twelve years plus," I have seen children as young as six flying small drones. The temptation to connect the battery and fly it in the back garden seems to be the norm—which I find very worrisome.

OBEY THE LAW

I really cannot stress this point enough: so many people purchase and fly drones without any thought of the potential consequences. It is this recklessness that gives drone flying a bad name and incurs even more strictures on those that fly safely. You will find a model aircraft organization in most countries; in the UK it's the BMFA (bmfa.org), and in the US it's the AMA (modelaircraft.org). You will find them online, so make sure to review their safety codes. Working with the FAA and other government organizations, the AMA has established (and continues to update) rules for UAVs and for FPV flight. Get involved and support your country's model aircraft organizations—and help protect our right to fly.

Countries all over the world are battling for a solution to the rise of commercial and hobbyist drones, as no one really knows how to address the situation. In Spain, where I sit and write this book, the government has put a total blanket ban on flying drones by anyone, anywhere, unless at a sanctioned radio-controlled airfield. One of the prevailing factors is the speed at which the drone industry is advancing. Technology in all areas is progressing quickly, which is pushing drone manufacturers to produce more and better products. Flight platforms are now stable, navigation ever more complex, camera systems have reached 4K definition, while battery life allows for much longer range and duration.

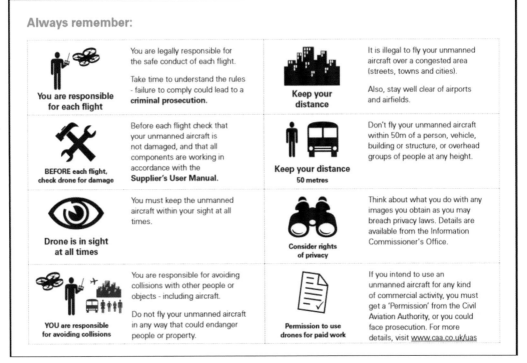

Always remember:

You are responsible
for each flight

You are legally responsible for the safe conduct of each flight.

Take time to understand the rules - failure to comply could lead to a **criminal prosecution.**

Keep your distance

It is illegal to fly your unmanned aircraft over a congested area (streets, towns and cities).

Also, stay well clear of airports and airfields.

BEFORE each flight, check drone for damage

Before each flight check that your unmanned aircraft is not damaged, and that all components are working in accordance with the **Supplier's User Manual.**

Keep your distance
50 metres

Don't fly your unmanned aircraft within 50m of a person, vehicle, building or structure, or overhead groups of people at any height.

Drone is in sight at all times

You must keep the unmanned aircraft within your sight at all times.

Consider rights of privacy

Think about what you do with any images you obtain as you may breach privacy laws. Details are available from the Information Commissioner's Office.

YOU are responsible for avoiding collisions

You are responsible for avoiding collisions with other people or objects - including aircraft.

Do not fly your unmanned aircraft in any way that could endanger people or property.

Permission to use drones for paid work

If you intend to use an unmanned aircraft for any kind of commercial activity, you must get a 'Permission' from the Civil Aviation Authority, or you could face prosecution. For more details, visit www.caa.co.uk/uas

CAA safety leaflet.

Additionally, we all need to understand the role of organizations like the Federal Aviation Administration (FAA) in the US and the Civil Aviation Authority (CAA) in the UK; they are there to prevent accidents from happening in their airspace. If some untrained owner flies a drone over an airfield and endangers a manned commercial flight, that is FAA business. But they are not there to receive complaints about the same person flying their drone over your backyard and watching you mow the lawn. That is a police or civil matter.

In the UK, the CAA have taken the initiative and issued a leaflet with all drones sold. This is a simple but much needed response to the growing number of hobbyist drones being sold. The leaflet simply points out certain rules that should be adhered to. If you have a drone and do not have the CAA leaflet, you can download it from caa.co.uk/Consumers/Model-aircraft-and-drones/Flying-drones/

Crashes can happen because of pilot error or hardware or software malfunctions. If you are flying anywhere near other people, you are putting them at risk! Even a small, insignificant quad-copter has fast-spinning propellers that could easily cause damage to a human. Your first priority must be the safety of others. Always make sure you maintain a safe distance between yourself, your drone, and spectators, no matter the size or whether flying indoors or outside.

You are not allowed to fly within 5 miles (800 m) of an airport unless you have permission from air traffic control, and they will normally only provide this to

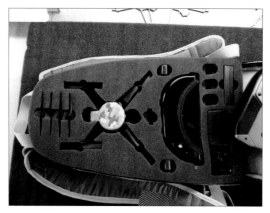

A drone case helps protect your investment and makes sure nothing gets accidentally damaged or broken.

commercial operators with an approved license. Where you are flying and the circumstances will require that you formulate your own strategy of what is a "safe distance" from people and property. When learning to fly, I recommend that you do not fly more than 100 feet (30 m) from your takeoff position. Constantly check that no one gets between you and your quad-copter. If anyone does enter the fly zone, it is your responsibility to act: land and wait until the area is clear before resuming flying.

Always remember that your quad-copter is a finely balanced robotic aerial machine; the slightest damage can cause serious problems, so make sure to always treat it with respect. You will have to store your quad-copter and GCS equipment as well as transport them to an airfield. To help you keep your drone safe and in good working order, you should purchase or make a storage/transit container. Although this will be an added cost, it will secure your drone and extend its working life.

SOME DO'S AND DON'TS

I am probably being a bit repetitive here, but when you are controlling a drone in the air for the first time, you really need to heed these important points—and I will hammer them home throughout this book:

- Once at the flying field do not rush to get your drone in the air.
- Make sure all your batteries are fully charged before you leave home or have the ability to recharge at the airfield.
- Check that there are no loose parts on the quad-copter.
- Fit the propellers securely making sure they have no chips or damage.
- Check the area and weather where you intend to fly.
- Check that the battery and additional components are secure. (It is not uncommon to see the battery fall from the drone by the connector.)
- Check if your quad-copter needs to be calibrated.

- When the battery is connected to the drone, check for satellite lock. DO NOT take off before that or you will not have a RTL available.
- Make sure the throttle (left stick) is all the way down and all switches are in the "back" position.
- Keep a safe distance between you and your drone, 10 feet (3 m) minimum.
- Always try to face the rear of the quad-copter when learning to fly as it helps with orientation.
- When flying, make sure you can always see your quad-copter.
- Know the flight communications range of your system; never fly beyond the capabilities of the system.
- Always assume the motors are armed when a battery is connected.
- After landing, the first thing you should do is push the safety switch if fitted and disconnect your battery.
- *Do not* turn off the transmitter until after you have disconnected the battery.
- Do not attempt to fly longer than your batteries' safe capacity; it is very bad for the batteries and can cause a crash.
- When learning, always use Stabilized mode for takeoff and landing. It's OK after several test flights to practice switching between AltHold and Loiter, but for safe manual control Stabilized mode is the best mode for recovery.

FIRST FLIGHT (ASSUMING RC TRANSMITTER IS IN MODE 2)

When flying a new model or taking your first flying lesson, always do so on a calm day when there is little to no wind. Before you take your first flight, there are several things you will want to familiarize yourself with. Imagine your drone is placed in front of you at a respectable safe distance and you are holding the RC in your hands. Do nothing until you are sure you have at least six satellites, though eight are better. It is normal to feel a bit apprehensive, but the key is to relax. This is not a toy—it is a flying robot. While it requires respect, you should not fear it. Carry out the following steps:

- Turn on your RC transmitter.
- Connect your battery to the quad-copter. (The lights will flash red and blue as the gyros are calibrated.)
- The autopilot will do pre-arm checks—if there is an error, the large LED light will flash yellow (see Chapter 4).
- Check that your flight mode switch is set to Stabilize.
- Press the flashing Safety switch until it turns solid.
- Make sure you have a good GPS signal—the RGB LED will blink green.
- Step away to a safe distance and arm the motors by holding the throttle down, and rudder right for 5 seconds.
- Raise the throttle to take off.

> **Author's Note:** Its takes around 5 seconds for the barometer and gyros to re-initialize. Do not hold longer than 5 seconds as you may trigger the Auto Trim function (see below).

JOYSTICK RESPONSE TEST

Slowly push the left-hand stick (throttle) up to provide sufficient thrust so that the drone almost takes off (wobbles a little). Then push the throttle to both left and right a little and see if the drone responds by leaning slightly in the same direction as your joystick movement. Do the same for the right-hand joystick and make sure the drone is responding as it should. Once you are satisfied, increase the throttle steadily but in a positive manner so that the drone breaks contact with the ground and flies into the air about 6½ to 10 feet (2 to 3 m) in height. Hold and hover at this position.

> **Author's Note:** Please remember the software we are running with our Pixhawk is written with open source community-created code—while it is excellent, unexpected results may happen. Should your drone fail to arm or takes off at a weird angle (or flips), then you have a problem. Disarm by pulling the throttle completely down and push to the left. Wait for the propellers to stop turning before approaching the drone. Press the Safety button on the drone until it blinks and disconnect the battery before switching off the RC transmitter.

MANUAL LANDING PROCEDURE

- When learning it is best to land in Stabilized mode.
- Fly the drone back slowly towards your selected landing site and hover at around 10 feet (3 m).
- Slowly reduce the throttle and bring the drone down to the ground.
- Once the drone is on the ground, hold the throttle down and push it to the left (disarm) for 10 seconds, and the drone will automatically disarm.
- Wait until the propellers have stopped before approaching.
- Press the Safety button on the drone.
- Disconnect the battery from the drone.
- Switch off the RC.

DRIFT

Assuming your drone took off and is hovering nicely, you should hold the throttle at its present position, release the other joystick, and check stability. Although your drone has the ability to self-stabilize in the air, it is highly unlikely that it will fly perfectly the first time and may tend to drift in one or more directions. The perfect scenario would be for the drone to take off and stay exactly where you want it with hands off the joysticks, so that it only moves when commanded to do so. Given that we have calibrated our drone as instructed, it should do its best to fly level. Another factor that makes the drone drift is balance. Your drone may be heavier on one side than the other, often the result of poor battery placement. Although you can correct this drift with the joysticks, it is best to trim the drone using the trim sliders on your RC transmitter. Correct one direction at a time. For example, if your drone is drifting to the left, counter this by pushing the rudder trim slider in the opposite direction to the drift. If the drift is quite significant, then you should adjust the appropriate trim slider to compensate. If the drift is less, then less trim should be applied, until the drone is stable in one spot.

> **Author's Note:** The chances of your drone staying exactly in one place are highly unlikely. Your drone is floating, and although the instruments will do their best, there is always a little movement.

SAVE TRIM AND AUTOTRIM

While it is almost impossible to get your drone to stay absolutely still in the air—even on a perfectly calm day—there are two methods of trimming your drone so that does not wander off all the time. Your drone will be trimmed to a certain degree when you go through the Wizard and carry out the acceleration calibration—you can also to this at any time using Mission Planner (Initial Setup / Mandatory Hardware / Accel Calibration).

Once you have trimmed your drone, it is possible to save your RC transmitter trims to the autopilot. Using Mission Planner, you will need to check that your Channel 7 switch goes above 1800 (Initial Setup / Mandatory Hardware / Radio Calibration). Next switch to (Config/Tuning / Extended Tuning) and set Channel 7 to "Save trim" from the drop-down menu—click Write Params.

- With your Channel 7 switch in the off position, fly your drone in Stabilized mode using your roll and pitch trims to get the drone flying level.
- Land, put your throttle to zero, and release the right stick (roll and pitch) to center.
- Switch Channel 7 to high for at least 1 second.
- Reset your roll and pitch trim to center and test fly again. If the drone is still not level, go through the process again.

Auto Trim

A much simpler way to trim is to do so automatically while in flight. You will need to do this on a windless day to get accurate results.

As with a normal arm, you push the throttle down and to the right; to Auto Trim, you do the same, but in this case hold it to the right for at least 15 seconds until you see the light flash red/blue/yellow in cyclic pattern.

- Take off and fly your drone in Stabilized mode for around 25 seconds, keeping it in the hover.
- Land, put the throttle to zero, and wait for about 5 seconds for the trims to be saved.
- Take off again in Stabilized mode and check how the drone is hovering—if you still have drift, repeat the process.

SIMULATORS

The growth in quad-copter simulators has not kept pace with the manufacturing of hobbyist drones, but there has been some progress. If I am to be honest, before writing this book I had little time for simulators other than Microsoft Flight Simulator X, which is an old copy I picked up at a Sunday market stall. At first I thought it was not serious enough to be of any use; then one day I sat down, set it up properly, and went through all the procedures from taking off to landing. Takeoff is no problem, but landing and lining a fixed-wing aircraft up with the runway, at the correct height, approach speed, and path taught me a lot. It was only at this stage that I began to realize that there was quite a lot of merit in flight simulators. The trouble was that quad-copters are new, and most of the good simulators don't really provide the operator with the right "feel" of a quad-copter in flight. The simulated drone seems to roll and do things a real drone would not, and they do not respond to the controls as they would if they were real. They lack the drift effect found in many rotary drones. Additionally, the graphics on most simulators are quite poor, which can lead to some degree of boredom. That said, for the complete beginner to learn the basics such as altitude hold maneuver, avoiding objects, learning how to keep orientation, and hand-eye coordination, a simulator is definitely a good idea.

Although a quad-copter is the easiest drone to fly, most beginners do not have the skills and reflexes to keep a drone in the air and fly it the way they would like. Understanding the drone and its deviations (high wind or too much speed) in flight generally causes newbies to overreact. This is normal, as they do not yet possess the flying skills, and they often end up crashing the drone. If it's a cheap toy then no matter, but most of the better hobbyist drones cost upwards of a $1,000, especially with a camera attached; that's a lot of money to lose. Trust me, its heart-rending to see it crash, and you can kick yourself for being so stupid. The only consolation is that it happens to the best of us, including me.

We all need time to develop our reflexes and skills so that when we fly we feel confident and comfortable with the drone. No matter how much your drone costs or how advanced your model may be, the skill with which you operate it is the single greatest factor in determining your success and safety. When things go wrong, ability and skill are the only help you can rely on.

Author's Note: My friend Pedro from Portugal is a drone instructor. He flies real aircraft as well as drones and has a lot of skill. Yet even he will tell you that there is no substitute for flying as often as you can—for Pedro it's every day. As with Tim Whitcombe in the UK, who is a senior examiner for the BMA, they both possess a deep understanding of flight and have achieved a level of confidence no matter the type of aircraft they are flying. They have reached this level only through years of practice.

You will want to know exactly what to look for in a simulator before you buy one, so choose it with care and deliberation. For what it's worth, I purchased several simulators and tested them and found the following models beneficial. However, this is my personal choice and not necessarily a recommendation.

ImmersionRC Liftoff FPV Flight Sim

Designed primarily for FPV racing, this brilliant new simulator allows you to use goggles that translate the rush of FPV drone racing to the digital world. Liftoff was developed by the game developer LuGus Studios and drone manufacturers ImmersionRC and FatShark. The simulator allows you to change motors, propellers, and batteries to help create your own drone. The graphics are suburb—especially the flight through the forest. While being designed purely for FPV racing, it makes a great simulation for beginners, although it is a little advanced.

Real Flight RF 7.5

RealFlight has historically been a tool for pilots of RC helicopters and fixed-wing airplanes. Multi-rotor models were introduced only in the most recent versions of the program. It's quite expensive at $130 and requires you to purchase the CD together with an InterLink Elite controller, but you can also use an RC transmitter you already own and are familiar with. I have not tested this simulator, as when I tried to connect to the website I got nothing. However, the graphics look good and there are now some fourteen model drones to choose from. As well as flight training, it also has a lot of instruction on aerial camera operations so you can also master the basics in gimbal control. You can also practice taking pictures under a variety of conditions such as day/night or in changing wind.

Phoenix Model Flight Sim

The graphics on Phoenix are really stunning, and there is very realistic 3D terrain mapping. It is a great simulator for both professionals and beginners. The scenes include an indoor gymnasium, a marina, plus a normal RC flying field, and a lot of extra scenes can be downloaded from the website. Phoenix version 5 features definitively refined physics plus new companion software you can use to create your own scenes and models—and lets you share them online. What is particularly good is the feel of the controls, which are extremely accurate. There is a free Phoenix "rolling" demo using the same graphics and physics engine as the full version of the software, which lets you see how Phoenix will look and perform on your system. Worth trying before you invest $129.

FPV Freerider

FPV Freerider

With a price tag of just $4.99, this is a really great simulator. Once again, it seems to be designed more for practicing drone racing than anything else, but all in all it responds and resembles a real drone in flight. I downloaded the free version on Android apps first, but this only gives you the desert scene and is fairly limited, so I tried the $4.99 download. I flew and crashed several time before I got the hang of it. It is selectable between FPV and LOS flying and has a self-leveling and Acro mode. There are also a lot of custom settings for input rates, camera, and physics. I did find the frame rate a bit slow, but you can change this by opting for a lower resolution.

Mode 2 is default, but the touchscreen controls support Modes 1, 2, 3, and 4. While it has the feeling of a game, it is a really good RC flight simulator. There are no objectives except the joy of flying. At first you will find that the controls are hard to handle, but that's because it's made to mimic real live flying.

Aerosim RC

Aerosim includes a CD with the software and the USB interface cable that you connect to your own transmitter. It supports many of the most common transmitters such as Spektrum, Graupner, Turnigy, and FreeSky, but I like it because I can use my favorite Walkera DEVO. Aerosim RC has done some recent updates and now includes a simulated DJI Phantom, but you cannot use the Phantom RC because it has no training output. I purchased my version many years ago and have used it frequently both for hobby and for training sessions. Its features are vast, and the best thing to do is try it; you get two minutes of play free of charge and after that you need to purchase a special USB cable. I like the idea that you can download and make your own map area using Google Maps. Cost is around the $80 mark, depending on where you buy.

Aerosim RC

Heli X Sim

This is a new one to me, but the main developer of Heli X holds a PhD in applied mathematics and is an expert in computational fluid dynamics. Like many RC pilots, he also holds a private pilot license, which means he really understands flying. Heli X has just been upgraded to version 6.0. There is a free trial version with the option to upgrade should you choose. Although designed mostly for helicopters, there are several popular drone models you can select. I have tried this out using the DJI Phantom and found it particularly good for orientation. Try flying the Phantom out until it is just a speck, then bring it back home. Look which way the speck moves in response to stick movement. The simulator also highlights motions such as ground effect, stall, and transitional lift. The training sessions include important features such as reaction.

FLIGHT LESSONS

It is always best to have the drone pointing away from you and keep it in this aspect as much as possible until you have gained some flight experience. It is easy to lose orientation, so if you know you are facing the rear of the drone, you will have better control.

When you first arm your drone and are ready to take off, first check that the drone is responding to the controller by doing the joystick response test.

TAKEOFF

By way of a reminder, you should never take-off until you have a GPS fix of at least six satellites showing on your GCS. The drone needs to know its take-off position if it is to return should there be a fault, or if "return to home" is deliberately invoked. Mainly depending to the amount of money spent on development, drones react in different ways. For example, it is possible to take off with the DJI Phantom by simply pushing the throttle stick slowly forward—it will rise into the air. Other drones, especially self-made drones, sometime require a bit of throttle power and joystick dexterity to lift them safely into the air.

The difference in performance is down to the individual drone settings. Since we are using a Pixhawk autopilot, we have the luxury of several modes; these include Stabilize, AltHold, and Loiter. It is possible to take off in any of these modes, although it is standard practice to take off in Stabilize mode first and then switch to either AltHold or Loiter once in the air. By comparison, most RTF drones will not take off until they have a set number of satellites and are immediately in what looks like Loiter mode; they hold their position and height the moment the throttle is released.

If your drone on takeoff simply makes a lot of noise and shuttles around on the ground, either the drone is too heavy to take off or your battery is low. If upon takeoff it leans in one or two directions off center (takeoff point), you should first try adjusting the RC trim to counter the movement.

Author's Note: In the first instance, it is best to takeoff with a steady, continuous upward push of the throttle until the drone is safely in the air. That said, over the years I have found a small trick for takeoff: by pushing the throttle smoothly forward to almost full power for a second until the drone leaps into the air, then immediately cutting the power in half, you will have a clean takeoff. It takes a little practice, and you need to know your RC throttle settings but, once you get the feel of it, it works great. I would also recommend that you use plastic propellers while learning as opposed to carbon fiber propellers, which are much more expensive.

Now that we are in the air and the drone is stable, it's time to do a little flying. With the drone in the air about 10 feet (3 m) high and about 2 feet (5/8 m) away from you, move both joysticks one at a time to see the drone respond, that is, rock it in the sky slightly forwards and backwards, side to side, and rotate it left and right. Now position it with the front of the drone facing away from you and start the following lesson. However—and yes I know I am being repetitive—heed the safety tips first;

Flying Tips for Beginners
- Always fly in a dedicated open area—do *not* fly in your backyard or any built-up area.
- Do a visual sweep of your flying area for people, objects, and hazards. Give buildings and trees a wide berth.
- There should not be another person other than an experienced pilot in support within 150 feet (50 m).
- Make sure your drone and RC or GCS has Stabilize, AltHold, Loiter, and RTL modes working correctly.
- Install all fail-safe functions available on your system, especially lost communications and GPS loss.
- GeoFence your flying area if the function is available to you.
- Ground test control before takeoff.
- Take-off and land in Stabilize when learning.
- Weather conditions may be different the higher you fly; take care of winds and sudden gusts.
- If the wind is stopping your drone from moving forward—land and abandon flying for the day.
- Take off and fly with your drone more or less stable on the horizontal plane without any control inputs.
- Avoid sudden or extreme transmitter control stick deflections; don't fight it—fly it.

- If you are fighting for control of your drone, land and fix it; something is not right—a hardware adjustment or software calibration may be required.
- Move the joysticks in small, precise movements—don't overcompensate or make wide movements.
- Control the throttle because your drone can climb or descend very rapidly.
- Quad-copters are particularly unstable in a rapid descent—use caution when descending, with small reductions in throttle.
- While learning, always keep the drone close: 100 feet (30 m) and no more than 15 feet (5 m) high so you can see its orientation.
- When first learning to fly, make sure you keep visual orientation of the drone—keep the rear towards you wherever possible.
- If you lose orientation, stop and place the drone in Loiter until you can assess its direction.
- In an open area, move toward the drone so you can physically see its orientation—if this is not possible, gently land the drone if it is safe to do so.
- If the drone has disappeared from view completely (flyaway), then activate the RTL function.

Crash recovery using a towel to prevent any injury should the motors be active or suddenly restart. Try to pick it up with your hands out of range of the propellers.

- When learning to fly avoid the temptation to fly too fast or too high.
- Wait until you have acquired considerable confidence in manual flight before progressing to automatic mode.
- Never fly longer than the batteries' safe capacity; it's not good for batteries and a sudden lack of power could cause a crash.
- Finally, it is better to land (safely) or stop flying your drone than to cause injury to a person; value safety above the cost of your drone.

CRASH RECOVERY

Normally if the drone has crashed because of pilot error, either into the ground or trees, the motors should stop running after a few seconds. In all cases where the drone has crashed, you should immediately pull the throttle fully down and to the left (disarm), but never assume that the drone has complied with your instructions. *Do not switch the RC off*, but make sure the throttle remains fully down.

Approach the drone with caution and throw something—like a large towel or jacket—over the drone. When picking the drone up, do so in a manner that keeps the propellers away from your body before removing the battery. A small drone recovery kit should consist of an old towel, small fire extinguisher, and a first aid kit.

FLIGHT TRAINING SKILLS

In this section I have laid out a set of flight training skills which we use in our company for beginners. You will find that both using a simulator and actually doing these movements should really help build your flying skills. As with flying a real aircraft, takeoff and landing are the two main skills; if you can take off well, stay in control, and land well, then you are halfway to becoming a good pilot.

End of test. Land the drone and disarm. Wait until the drone has landed and motors have stopped before approaching.

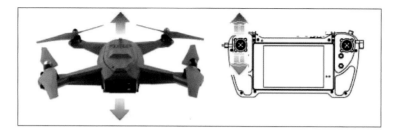

Take off and land with the back of the drone toward you.

Hover in the same place keeping the back of the drone toward you/check slight movements of forward/back/left/right.

Hover in the same place keeping the back of the drone toward you and rotate the drone left /right.

Slowly fly forwards/backwards /left /right with the back of the drone pointing toward you.

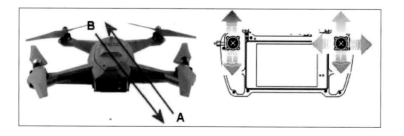

Pick out a clear landing place around 30 feet (10 m) away. Fly to it and land. Take off using the direction in B, and return to original takeoff point (A).

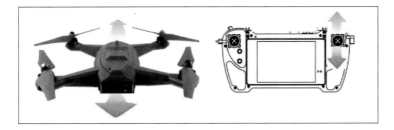

Fly to a spot 100 feet (30 m) away and hover over the spot. Then return to takeoff point.

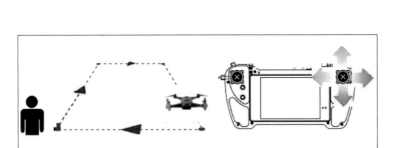

Fly a clockwise 100-foot (30 m) box with the front of the drone pointing forward and returning to your takeoff point.

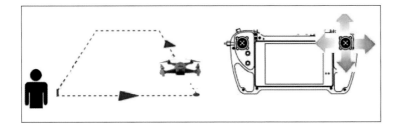

Fly the same box in reverse (counter-clockwise); this time wait at every corner before proceeding.

Fly to the hover at 10 feet (3 m) and then rotate the drone 360 degrees clockwise.

Fly to the hover at 10 feet (3 m) and then rotate the drone 360 degrees counter-clockwise.

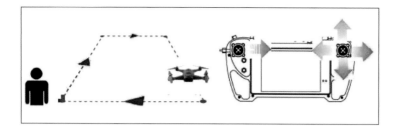

Fly to the hover about 15 feet (5 m) and then turn the drone 90 degrees to the right and fly a four-point box. Stop and hover at each turning point.

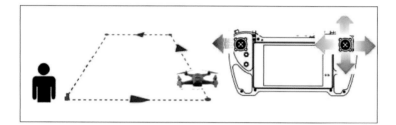

Repeat the preceding exercise with the drone flying counter-clockwise.

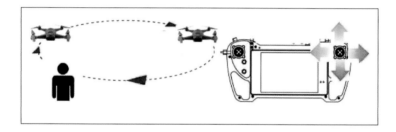

Start with the drone pointing forward and fly a uniform clockwise circle maintaining an even height.

Take off and hover at around 10 feet (3 m). Rotate the drone until the drone is looking at you. Hold the position as steady as possible.

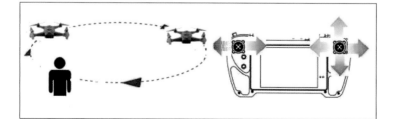

Fly a clockwise circle about 15 feet (5 m) high, keeping the drone looking at you all the time. Maintain height and uniform.

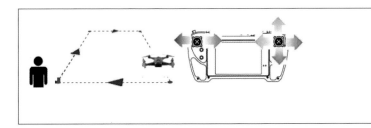

Fly a clockwise 100-foot (30 m) box with the front of the drone pointing at your position at all times.

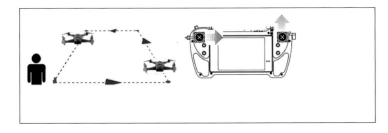

Fly a clockwise 100-foot (30 m) box and rotate the drone in the direction of travel.

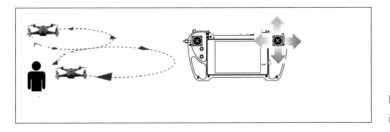

Fly a clockwise circle and keep the drone in the direction of travel. Maintain height and uniform.

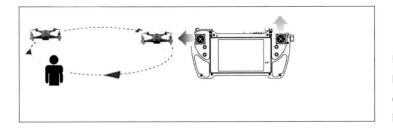

Fly a clockwise circle and keep the drone pointed towards the center. Maintain height and uniform.

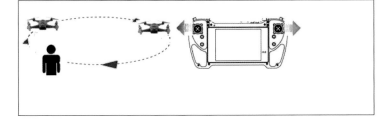

Fly a figure eight, maintaining height and uniform.

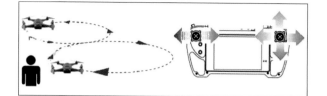

Fly a figure eight, with the drone pointing in the direction of travel.

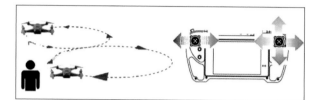

Fly a figure eight, with the drone always pointing towards the center.

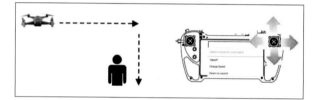

Fly the drone 100 feet (30 m) away and press the Return to Launch button.

TURNING

There are two ways of turning your quad-copter: one is called a yaw turn, and the other is a bank turn. A yaw turn uses a combination of the left and right joysticks—throttle to maintain height, yaw to turn, and pitch to advance the drone through the turn. With a bank turn, you maintain throttle height while using the right joystick to roll the drone while advancing the drone through the turn. A bank turn is the easiest turn to practice for a beginner because it is a little slower to achieve.

QUALIFICATIONS

The FAA states that you become part of the US aviation system when you fly a drone in the nation's airspace. Model aircraft operations are for hobby or recreational purposes only. The FAA has promoted "Know Before You Fly," a campaign to educate the public about using unmanned aircraft safely and responsibly. Individuals flying as a hobby or for recreation are strongly encouraged to follow safety guidelines, which include:

- Fly below 400 feet (120 m) and remain clear of surrounding obstacles.
- Keep the aircraft within visual line of sight at all times.

- Remain well clear of and do not interfere with manned aircraft operations.
- Don't fly within 5 miles (8 kilometers) of an airport unless you contact the airport and control tower before flying.
- Don't fly near people or stadiums.
- Don't fly an aircraft that weighs more than 55 pounds (25 kilograms).
- Don't be careless or reckless with your unmanned aircraft—you could be fined for endangering people or other aircraft.

The statutory parameters of model aircraft operation are outlined in Section 336 of Public Law 112–95 (the FAA Modernization and Reform Act of 2012). Individuals who fly within the scope of these parameters do not require permission to operate their UAS; any flight outside these parameters (including any non-hobby, non-recreational operation) requires FAA authorization. For example, using a UAS to take photos for your personal use is recreational; using the same device to take photographs or videos for compensation or sale to another individual would be considered a non-recreational operation.

As I finish this book, the FAA announced new rules for everyone who owns or wants to fly a drone. The system is one of registration using a web-based aircraft registration process for owners of small unmanned aircraft weighing more than half a pound (250 grams) and less than 55 pounds (25 kilograms), including payloads such as onboard cameras.

The registration began on December 21, 2015, and costs $5, but the first 30 days were free in an effort to encourage as many people as possible to register quickly. Registration is a statutory requirement that applies to all aircrafts. Under this rule, any owner of a small drone who has previously operated an unmanned aircraft exclusively as a model aircraft prior to December 21, 2015, must have registered no later than February 19, 2016. Anyone who purchases a model aircraft after December 21, 2015, must register before their first flight outdoors. In my opinion, this is certainly a step in the right direction. I believe that registration involves getting a unique number that must be attached to the drone for identification should an incident occur. To register, you need to be over thirteen years old or have someone meeting the required age to register for you—go to: faa.gov/uas/getting_started/.

The law for hobbyists in the UK is basically the same as in the US and in both countries any commercial use requires permission from either the FAA or CAA. The UK is addressing drone safety by pursuing a path of education. In December 2015, a joint British Model Flying Association (BMFA) and Civil Aviation Authority (CAA) initiative was launched at the Drone Show NEC in Birmingham. It is hoped that the initiative will encourage clubs and individuals to participate in a new voluntary proficiency certificate.

TRAINING FOR COMMERCIAL LICENSE

Drones are increasingly being used commercially in a wide variety of employment, such as security, geospatial mapping and surveys, construction, property management, environmental monitoring, incident management support, aerial photography, and event filming to name a few. Those people who are interested in using drones for commercial purposes must first obtain a CAA-recognized qualification in the UK, as well as seek special permission depending on the weight of the drone (45 pounds/20 kilograms max) used and where it will be used. There are several companies that offer a CAA-recognized National Qualified Entity for Small Unmanned Aircraft (NQE SUA).

For those in the US who want to earn money as a professional drone pilot, the FAA requires you and your business to apply for and successfully obtain a 333 exemption. By law, any aircraft operation in the national airspace requires a certificated and registered aircraft, a licensed pilot, and operational approval. Section 333 of the FAA Modernization and Reform Act of 2012 (FMRA) grants the Secretary of Transportation the authority to determine whether an airworthiness certificate is required for a UAS to operate safely in the National Airspace System (NAS). This authority is being leveraged to grant case-by-case authorization for certain unmanned aircraft to perform commercial operations prior to the finalization of the Small UAS Rule, which will be the primary method for authorizing small UAS operations once it is complete. It took a while for the FAA to respond to the demand, but it now looks like the agency understands the demand and what skills are required, as I believe that during 2015, more than 2,000 approved exemptions were made.

At present, if you plan to fly your drone recreationally, you do not require certification or exemption. This will change in the near future as both the US and the UK are looking at some form of basic flight certification for any drone flying.

In both cases the basic rules are the same:
- Don't fly above 400 feet above ground level.
- Don't fly within 5 miles (3 in the US) of an airport/landing strip.
- Do not fly at night.
- Keep your drone within line of sight
- Don't fly in NOAA zones and obey all TFRs/FRZs (Temporary Flight Restrictions/Flight Restricted Zones)
- Fly safely near pedestrians, wildlife (no flying over national parks), built-up areas, and so on.

Author's Note: The preceding text is adapted from the FAA and CAA websites for model aircraft operations.

SUMMARY

Flying a drone is the best part of this hobby; while building and calibration are great to learn, there is no substitute for seeing your drone lift off and fly around the sky. For all those people interested in flying model aircraft, helicopters, or drones, there is nothing better than to plan a visit to your local flying club. Check the Internet to find your local flying field, make your way down there, and simply talk to them. There will be some who enjoy building model aircraft to scale, while letting others fly them; many are experts at flying helicopters, and more recently there will be some flying drones. You can also find out if the club has an instructor so you can learn how to fly properly; at this point, you will get a lot of advice on what to buy and what not to buy. The advantages of joining a model flying club are many, as you gain from years of experience. Even those who are a little nervous can get help via a *buddy lead*, which simply means both student and instructor have an RC transmitter that is interconnected so that the instructor can take over at any time (similar to mechanisms in place when learning to drive a car). You will also learn from the ground up all the safety requirements.

Take heed of the flying lesson in this chapter, as they have been designed to guide you through a set of steps, each of which offers a small improvement. Although it is true that many drones will fly straight out of the box, your drone's lifespan will be dramatically shortened if you don't learn to fly it properly. Finally, take note of the regulations in your country and heed them. Each violation is a blemish on the drone industry and only serves to curtail the activities of those who fly within the law.

Chapter 7

DROIDPLANNER

first saw DroidPlanner mid-2013, and immediately thought that it was a brilliant tablet application for automated flying. Designed by Arthur Benemann, it looks and feels like a younger sister to Mission Planner, which makes it easy to accept and use. Whereas Mission Planner is designed for a laptop or desktop, DroidPlanner is designed for a tablet. Over the past two years, DroidPlanner development has moved at an amazing pace and is now supported by 3DR, which in late 2014 took out the back end to 3DR Services and produced Tower, also known as DroidPlanner 3.

I have opted to stay with DroidPlanner 2 for various reasons, the main one being it is slightly easier for first-time users. It will work on most Android tablets. As a personal choice, I use a Nexus 7 tablet and have successfully flown most of my drones without a single glitch. It was this combination of great GCS software coupled with the extremely good daylight screen on the Nexus 7 that promoted the use of both in the development of the UXC-850 GCS (see Chapter 9).

There are several sets of GCS software available; some are designed for a specific drone while others are open source. This one from Walkera also has on-screen joysticks—which work extremely well, although they do require a light touch.

The most exciting thing about DroidPlanner is that, because it runs on a tablet, it can be used outdoors much more easily that Mission Planner, which requires a laptop. Additionally, it provides map caching while in the field and out of Wi-Fi range via the Google Maps API. You simply do this by zooming into the required flying area while connected to WiFi and then closing the DroidPlanner app; when you reload the app, the required maps reopen from the cache. But a note of caution: it does not always work. To overcome this you can use the Prefetch function in Mission Planner and transfer the required maps to your tablet; in this manner, you can save several different areas of mapping, allowing you to be sure that the map area you need is always available while the device has no network connectivity.

DroidPlanner lets you control your drone by simply touching the screen. You can take off and land automatically, and you can make, save, and fly way-points in a variety of combinations. You can select Follow Me and make your drone not just follow you but in a defined way, such as from behind or in front, or even circle you as you walk, run, or cycle. You can follow the drone's progress on the map and see what's happening via the HUD.

If it has a fault, it's the fact that there is no video downlink screen in DroidPlanner 2, but this has been addressed with the latest version of Tower (DroidPlanner 3). For the purpose of this book and to familiarize you with the basics, I am using version 2.8.6.

DroidPlanner screen with first version of integrated video downlink.

Author's Note: Although it is possible to fly your drone using only a tablet with Tower installed, as I have done on many occasions, you are strongly advised to have your RC controller running at the same time as a safety backup. Likewise, I have used Mission Planner on a 10-inch Windows-based tablet with a telemetry device attached with a normal USB cable, and it worked extremely well, although I still required a RC transmitter for control. Some drones such as the Walkera Scout X4 will fly extremely well using just a tablet, but again I would advise against doing this.

Although DroidPlanner is designed for advanced users and experienced pilots, there will probably come a time when you will want to give it a try. To this end there are a few do's and don'ts. The main one is to leave your drone settings alone; do not adjust any parameters unless you are an experienced pilot. Double-check that there are no other GCS systems working on your laptop or mobile phone. Only use missions made and saved in Tower. Double-check that your mission is uploaded.

CONNECTING TO YOUR DRONE

In order to establish a link between your drone and the DroidPlanner application, you need communications. This is accomplished by connecting a telemetry unit via an on-the-go (OTG) USB cable to the tablet. An OTG USB cable allows those devices to switch back and forth between the roles of host and client. The telemetry unit can then talk to the telemetry unit in the drone (transferring data). If your tablet is already switched on and running the moment you connect the USB cable, the drone will respond by asking if you want to connect to DroidPlanner. If you have more than one set of GCS software on your tablet, it will offer you a choice.

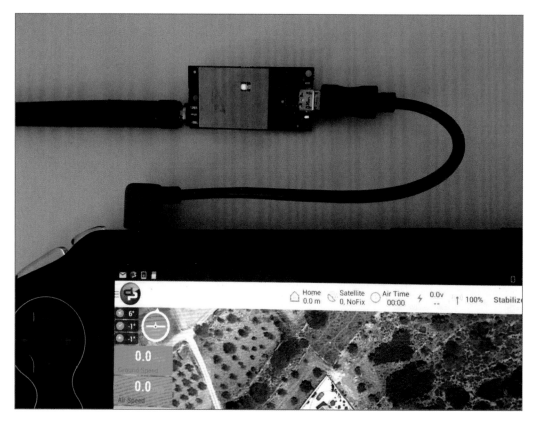

OTG USB cable used to attach telemetry to Nexus 7.

DROIDPLANNER SCREENS

The GCS software will always open on the operations page (Flight Data). This has a white strip at the top with a DP symbol and three horizontal bars sign in the top left-hand corner (a drop-down page menu) and three dots at the right hand corner which allows you to Connect/Disconnect to your drone as well as Send/Load missions and more. Most of the screen is taken up with the map, which, if you have active Internet or embedded maps, will populate. There are three buttons on the right of the map and a HUD on the left together with Ground Speed/Air Speed/Climb Rate/Altitude.

You should now make sure you have a fully charged battery attached to your drone and everything is functioning normally. Tap Connect on the DroidPlanner screen and wait for the telemetry to connect. You will see the top white strip suddenly populate—or if your telemetry is not working, an error message. Once connected you will see the telemetry data appear, such as how many satellites you have, flight time, communications strength, and flight mode. You will also see the Arm/Drone buttons which appear in the bottom center screen.

DroidPlanner Flight Data screen.

Tap the DP symbol, open the drop-down page menu, and select Editor. On this page you will see that the HUD has been replaced by four new symbols; these allow you to plan way-points or carry out actions.

Touching the first symbol (highlighted) allows you to input a single way-point. You will also see two other buttons that say Normal/Spline. If you place your finger over the way-point, you can adjust it to the required position.

Author's Note: The difference between a way-point and a spline way-point is that with a spline way-point, when executed, the vehicle will fly a smooth path both vertically and horizontally instead of going in a straight line and abruptly turning at each way-point as it lines up for the next.

Touching the paintbrush symbol allows you to draw a flight route on the map. Once your finger is removed, the route is populated automatically with way-points. Holding your finger on each way-point allows you to adjust its position.

Touching the third symbol allows you to use your finder to surround an area, which is then populated automatically with a mesh; this is either a search or survey polygon.

Touching the bin symbol will allow you to delete one or more way-points in any of the above procedures.

Touching any way-point created will open a new window where you can adjust the way-point's attributes. In the case of an area survey you can set the camera to adjust the Hatch Angle/Flight Altitude/Overlap/Sidelap.

With all normal or spline way-points you can set the altitude and any delay before proceeding to the next way-point. Each way-point can also be assigned to a function such as Circle/Structure Scan/Region of Interest.

For example, selecting a Structural Scan allows the pilot to circle a selected structure and climb up that structure at set interval heights; the camera will always point towards the center as the drone circles and climbs.

Select Region of Interest and then place your finger on any of the way-points (for example, way-point 4); this will allow you to move the way-point out of the flight path and reposition it. Once repositioned, the drone will fly the route with the camera always looking at the position of way-point number 4.

For most routes, we would set the first way-point as "Takeoff" and the last as "Land."

Author's Note: Once RTL has been activated, do not try to take control of the drone until after it has completed its RTL procedure, landed, and has the motors turned off. Interference in the RTL function can cause the drone to crash.

MISSIONS

Touching the three-dots symbol in the right-hand corner provides you with a drop-down menu. This is where you can Save/Open/Send/Load your mission. Before I fly using Auto, I like to double-check that the flight mission is what I intend. I do this by naming and saving the mission and then deleting the mission from the screen and reloading it. Once you are ready to fly your mission, you need to switch from Stabilize mode to Auto.

You will note that the list contains a lot of other modes, such as AltHold, Loiter, RTL, Circle, and Land. Be extremely careful when selecting your mode and make sure you press the correct one.

Saving and Loading Missions

DRONIE

You will have noticed that, when connected—alongside the Arm button—is the word "Dronie." A dronie is basically a selfie that you would ordinarily take using a selfie stick, but in this case is done by our drone. Before you put your drone into Auto, always make sure you have plenty of room when you make a dronie; set the drone at least 25 feet (8 m) away from you and check the surrounding area is clear up to 325 feet (100 m). Best to place your drone so that the camera is facing you.

Turn on your controller and keep the throttle stick down; this is in case of an emergency. When your copter and camera are ready, connect the radio to your Android

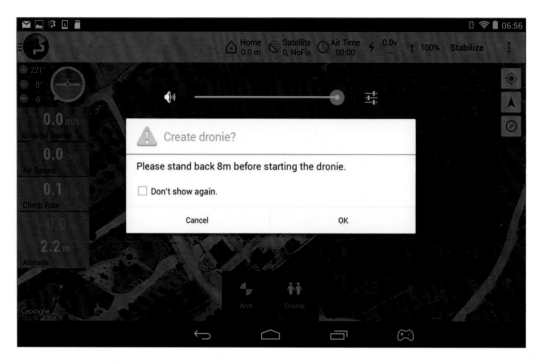

Dronie, the aerial selfie.

device and open DroidPlanner. Hit Connect to establish a connection. Press the Safety button on the drone until it is solid—step back at least 25 feet (8 m). Next you'll see two buttons: Arm and Dronie. Tap Dronie to set up the flight path. You will be warned if you do not have a good GPS fix. The map will update with the path plan and a warning to stand back from the drone. You will also hear "waypoint saved to drone"—select and tap Auto.

The drone will take off, turn the camera toward you, and fly back and up. (It's a bit scary the first couple of times.) When it reaches its peak at 165 feet (50 m) out and 100 feet (30 m) up, it will return to land and disarm. You can modify the Dronie as you would any other mission. Go into the Editor after tapping dronie and you should see a path you can modify. The first way-point is its takeoff height. The second point is the ROI (which should be where you're standing). The third way-point is the peak altitude; you can set this to be as high and as far out as you want. The fourth way-point is a slowdown point when it gets near you on descent.

All the functions in DroidPlanner are self-explanatory, and you are advised to master each one individually until you become proficient. DroidPlanner also allows for several other functions; again, these are self-explanatory.

SETTINGS

Settings allow the operator to change the basic functions of the DroidPlanner such as using offline maps which is found in User Interface.

PARAMETERS

You need to be connected in order to see the current Parameters. These can be modified and sent to the drone (expert only recommended). Parameters can also be saved in the folder file.

CHECKLIST

Checklist indicates the state of the GPS, drone battery, and more.

CALIBRATION

Calibration allows you to calibrate the drone in the field. Make sure you complete the calibration 100 percent.

FIRST AUTOMATIC FLIGHT

I would suggest that, for your first few automatics flights, you start by taking off in the normal manner using your RC transmitter in conjunction with DroidPlanner. Carry out the startup procedure as for a normal flight with your RC to the point prior to pressing the Safety button. At this stage it is best to set

Keep the drone pointing away from you and at least 10 feet (3 m) in the air in a stable hover.

aside your RC transmitter and concentrate on the tablet. With the DroidPlanner app running, press Connect and you should see all the telemetry data appear on your screen. Check the status of DroidPlanner to make sure it is in Stabilized mode.

Press the Safety button on the drone until it goes solid.

Tap Arm on DroidPlanner or use the RC transmitter, and you are now ready to fly.

Because DroidPlanner has no on-screen joysticks, you will need to use the RC transmitter if you want to take off in manual mode. Alternatively, you can fly in Automatic mode using only the tablet. When starting off, I would suggest you take off using your RC transmitter. Slowly release the throttle up to 50 percent, where it will center and remain. If the drone has not lifted off, increase the throttle until the drone is approximately 10 feet (3 m) in the air—then tap either Loiter or Altitude Hold and let the drone settle.

It is advisable to make sure the volume is turned all the way up and that the speech function in DroidPlanner is activated. This will aid your flying and keep you informed of the flight mode of the drone each time you change. Be aware that the voice can be a little annoying with lots of warnings that are not necessarily true. The modes in DroidPlanner provide the same functions as those in Mission Planner, but I elaborate in the next sections.

Stabilize Mode

Stabilize mode is the primary operating mode for flying and can be considered its manual flight mode. Stabilize automatically levels the quad-copter and maintains the current heading, while allowing the pilot full control over the throttle. Stabilize is good for general flying. The autopilot must always be set to Stabilize mode when you begin in order to be able to arm the ESCs before takeoff. It is *very* important to be able to easily and rapidly switch back to Stabilize mode from any other mode in order to regain control from any unexpected or undesirable flight behavior.

AltHold

AltHold means the throttle is automatically controlled to maintain the current altitude. Current altitude and movement is measured by the barometer in the autopilot. While in AltHold, the pilot still maintains control over roll, pitch, and yaw, which operate the same as in Stabilize mode. It's important to have the drone flying in stable condition before entering AltHold mode. While in AltHold mode, altitude can still be manually changed by raising or lowering the throttle control beyond a large central dead band. After using manual override to change altitude, it is important to reposition the throttle stick back to as near your normal hover position as possible, as AltHold uses the RC transmitter throttle stick center position as the center point for all manual altitude adjustments. When you switch back to Stabilize mode from AltHold, the drone will immediately go in the direction of any throttle stick displacement, potentially causing it to drop or rise at an alarming rate. Should the drone rise rapidly when you switch to AltHold, it may mean that it is suffering from too much vibration. Likewise, when landing in AltHold, the downwash may cause the drone to become erratic; this is due to the barometer being affected by pressure changes.

LOITER MODE

When switched on, Loiter mode automatically endeavors to maintain the current location, heading, and altitude. Wind, PIDs, and sensors will affect the effectiveness of maintaining position. The stronger the wind, the greater the location deviation. While in Loiter, the drone's location can be manually adjusted with the control sticks. As with AltHold, when you switch back to Stabilize mode from any altitude holding mode, the copter will immediately go in the direction of any throttle stick displacement, causing it to drop or rise at an alarming rate. Always make sure the drone is at a reasonable height and have your finger on the throttle when switching from Loiter to Stabilized mode.

RTL MODE (RETURN TO LAUNCH)

When you have acquired a good 3D GPS lock of six satellites or more, the takeoff point location is registered in the autopilot when the drone is armed. When RTL mode is selected, the drone will return to the takeoff location. By default, the drone will stop whatever it is doing, then first rise to at least 50 feet (15 m) before returning to the takeoff location, or it will maintain the current altitude if it is higher; this value can be changed on the configuration settings. It is important take into account any obstacles that may affect any RTL when you first approach your flying site and carry out a safety assessment. RTL is a GPS-dependent function, so it is essential that GPS lock is acquired before attempting to use this mode. When RTL is activated on any drone, you are advised to let it carry out the task, land, and let the motors stop turning. Any interruption in the RTL function before it's completed may cause problems or a crash.

AUTO MODE

Auto mode allows the drone to follow internal mission scripts that are stored in the autopilot to control its actions. Mission scripts can be a set of way-points or very complex actions such as takeoff, spin *x* times, take a picture, and so on. When an action, location, or set of way-points is created, you will be prompted to add such information as the height the drone should fly at as well as other criteria to keep the drone flying efficiently (see "Way-Points").

Auto mode is extremely good, but you are advised to keep a close eye on your drone while it is flying. Should anything go wrong, the pilot should immediately switch to another mode such as Loiter, which will interrupt the Auto mode and give the pilot control. Tapping Auto again means the drone will restart the mission from the first instruction. Until you have become a proficient flyer, you should always have a RTL as a final command to any Auto mission.

You can program the mission so that the drone takes off automatically, but be aware that your throttle must be at zero before you tap Auto, as the moment you raise the throttle the drone will start its mission.

Author's Note: Having flown many Auto missions, I have learned to respect the mode. While it works well, human error can cause serious problem, crashes, and flyaways.

- Before you fly, make sure your drone has no old Auto missions in its memory, because if you tap Auto by accident, the drone may attempt to fly the mission.
- When you make a new mission, upload it to the drone from your GCS, then delete it from your GCS, and reload it from the drone. This way you double-check that both the drone and the GCS have the same mission.
- When flying in Auto mode, always have an RC controller working in tandem.

There are many other modes in both Mission Planner and DroidPlanner that you can try when you have become a proficient pilot. They include Acro, Brake, Drift, Follow Me, and Sports mode. The best of these is Follow Me, which enables the drone to follow you at a specified high and distance. It can be programmed to follow you from the front, sides, rear, or even circle you during your activity. This is great for doing a selfie video during an activity sport such as cycling or even wind surfing.

MAPPING

First check that you have mapping for your area of operation. Where an Internet connection is found it is possible to use Google Maps which generate automatically when the program loads. In areas where there is no Internet access, you will

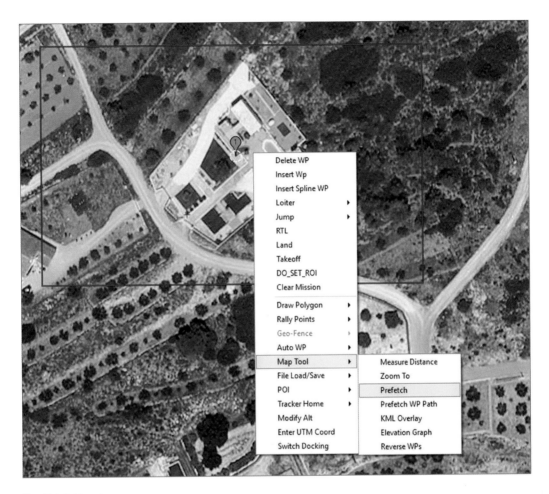

Pre-Fetch Mapping

need to go to Mission Planner, select an area using the Pre-fetch function, and select the area of operation.

Open Mission Planner and zoom into the area you desire, select Flight Plan, and check the best mapping in the map selection window under Action at the top right of the window. Press Ctrl and use the left mouse button to highlight the area you want to capture in blue.

Click the right mouse button to bring up the list of functions and select Prefetch from Map Tool. Click OK when it asks for zoom level, and the maps will start to download.

Author's Note: Do not make the capture area too large as it could take several hours. The size shown on the image above is average. I have downloaded around twenty different areas around the world where I have been flying. I rename each folder with the place name as a reminder.

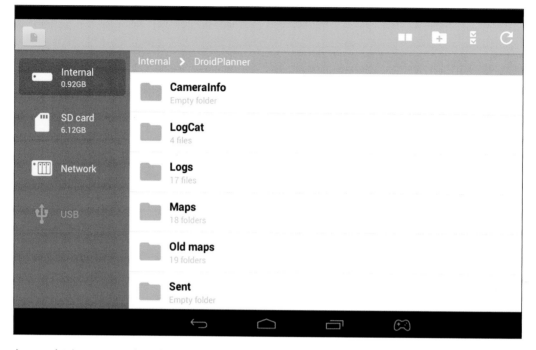

Any pre-fetch maps transferred to DroidPlanner must be renamed Maps.

You can change the subfolders under the Maps folder at any time. However, it is best to simply rename the folder according to the area it represented, such as FlyingField, for use later on. The file must be renamed Maps again each time an area is used. For example, you may have a folder with all the maps of your home flying field named FlyingField; to use this, you must rename it Maps so that DroidPlanner recognizes it.

SUMMARY

Learning to use DroidPlanner is a real pleasure, and seeing your first automatic flight take off and carry out your mission is a real joy. I have found using the DroidPlanner app has really opened up a lot of possibilities for me and broadened my love of flying. It also illustrates just how far drone flying has advanced. Once you have mastered the basics with DroidPlanner, I would suggest that you download and take a look at Tower (DroidPlanner 3), as this application has many new features, including downlink videos for some models.

I personally found it beneficial to fly my drone into the air before committing to Auto and sending my drone on a mission, as opposed to using auto takeoff and landing. Make small missions with one or two way-points and slowly build up your confidence with more complicated maneuvers. Although flying automatically can be exciting, always make sure you have your RC transmitter as backup in the event of your drone doing something it is not programmed to do.

The first time you fly in Auto mode, you may think the drone is flying way beyond what you originally though—this is a normal reaction, as it is difficult sometimes to estimate and compare what you plan on the tablet screen and the distance the drone has to fly. Remain calm and let the drone carry out its mission. You can interrupt most missions and take control manually, or you can invoke the RTL mode. Remember that when you press RLT, let the drone return. Do *not* interrupt until the drone has returned, landed, and the propellers have stopped turning. You might get away with it, but several times I have tried to stop an RTL, and it has caused the drone to crash.

Chapter 8

FIRST PERSON VIEW (FPV)

You will find a lot written about first person view (FPV), as it was one of the first additions to be made to the basic drone frame. When flying using FPV, the pilot has a vision from the cockpit perspective looking forward, straight down, or to the side depending on whether the camera is fixed or moveable. When the camera is pointing forward, it's as if the pilot were actually inside the drone flying it.

In many respects, FPV has driven the quad-copter industry improvements of the quality of flight, most especially in terms of stability and anti-vibration. And the range and quality of video transmitters have vastly improved. In essence, FPV is simple, putting a camera on a drone and transmitting the video feed live from the cameras to a visual means on the ground, normally at the GCS. The visual means can be either a small screen or a set of FPV goggles. Although this seems fairly straightforward, in the early days, the camera was fixed and suffered the vibrations—known as "jello"—which made the screen look wobbly. In a few short years this has completely changed; not only do we have super stable 4K cameras, but we can turn them in any direction required to get the appropriate shot. In the meantime, the transmit range has improved from a few hundred feet to well over 1¼ miles (2 kilometers).

At this stage I must point out that there is a vast difference between using the two viewing types. One allows you to look at a screen and also observe your drone and your surroundings. But, when wearing FPV goggles, you only see what the camera is seeing and are blind to everything else around you. The difference generates a vast divergence in the learning path of each viewing type.

If I am flying and observing the camera downlink on a monitor, I still have the ability to see my drone actually in the air (LOS). Additionally, I can also see where my drone is in relation to any obstacles or collision dangers. If I am using a touchpad GCS, then I am also making sure I am pressing the correct buttons in order to carry out my commands—changing flight modes, for example.

By comparison, if I am flying using FPV goggles, then my vision is restricted to the field of view provided by the camera lens. It is true that on-screen display (OSD) will provide me with telemetry data such as direction, height, position, and battery level, but this can also detract from the actual visuals of the camera. Unless you have two or more cameras positioned on your drone, your view is restricted. This means you may not know what is below or behind you. Moreover, your brain has to absorb two different perspectives when you are on the ground while the drone is in the air—two mental anomalies. Some argue that wearing FPV goggles is like being in the cockpit of the drone—it is not. Perception and depth of vision are extremely difficult until you get used to wearing goggles. Nevertheless, using FPV goggles has some advantages once you have mastered the technique.

On the other hand, FPV aircraft can be flown well beyond visual range, limited only by the range of the remote control and video transmitter. Where telemetry is used, the pilot has a set of instruments to observe just like a real pilot, the data from which is overlaid on the video screen. This provides the direction of flight and distance from home as well as keeps control over communications strength. Flying beyond LOS is allowed in some cases, but you will need a *spotter*, someone who can see the drone and what it's doing. You should also be aware that your FPV images will fade and stop long before your RC or telemetry (unless you are using a long-range digital video downlink system). This means you will then be reliant on flying your drone safely home or hitting RTL.

MAIN USES OF FPV

FPV is not new—the very concept of using a drone was to attach a camera for observation. For the military, almost all drones provide some form of FPV, as it is part of the guidance and confirmation system. In some cases it is a combination of satellite vision and that streamed from the drone's multiple camera systems. Most of the larger military drones are flown from GCS setups in mobile trailers or underground bunkers where the operators sit and drink coffee while the drones fly. Even the smaller military drones have excellent FPV, including the small Black Hornet and the larger Desert Hawk.

On the commercial front, some of the organizations that benefit from FPV include the police and those interested in aerial photography (such as the film industry). The advantages for "blue light" services range widely from security of an installation to crime prevention. The use of a drone for search and rescue

A Marine corporal controlling a Raven drone.

Photo courtesy of Corporal William Perkins

services has already proven to be of great benefit during the recent earthquake in Nepal. Having an aerial eye in the sky allows for so many different aspects in so many industries, using a thermal or near-infrared camera in the FPV mode to detect body heat or plant growth being a perfect example.

The film industry has been using aerial photography overseas for many years, but more recently the FAA has granted exemption to several companies. Flying-Cam Inc. won an Oscar for its rooftop chase scene (filmed in Istanbul) during the James Bond film *Skyfall*.

The hobbyist market has also found a new use for FPV: drone racing. The sport of drone racing is still in its infancy, but I personally believe that it will help secure a better safety record for drones. My reason for saying this is based purely on the corresponding improvement in pilot skills. Drone racing involves small, custom-built drones with an attached FPV camera that broadcasts video feeds back to specially-designed goggles or a monitor that allow pilots to see what the drones see. The drones used in this sport are extremely fast and can reach speeds of up to 70 mph.

Although some will take their drones to the local model airfield and fly them, for the most part I have found that drone pilots tend to fly alone. Drone racing is the start of something that will bring drone enthusiasts together. As the sport grows, it will help develop pilot skills, hand-eye coordination, and, who knows, somewhere down the line, an automated drone race. Imagine that: pre-programmed drones flying at speed through a forest, avoiding trees and bushes alike. This will involve developing rapid sense-and-avoid techniques, something that will advance drone technology to a safer level. Drone racing does have its problems, such as operating

Drone racing takes a lot of skill, but it is an exhilarating sport which will grow rapidly to the benefit of all.

in a dense environment, poor video reception, and the need for reliable communications. The great thing about competition is that pilots and engineers will endeavor to overcome these problems, which in turn will benefit all those who fly drones. There are several excellent racing drones available in ARF form, and building them is not too dissimilar to the build process covered in this book.

> **Author's Note:** The 2015 FatShark US National Drone Racing Championships took place at the California State Fair. There were 120 pilots, plus hundreds of volunteers and sponsoring organizations. And although you would expect the American Model Association (AMA) to be in attendance, you would have also found people from NASA and the FAA. The event was so large and popular that it was covered by CBS, FOX, ABC, and NBC television, making it the largest FPV racing event so far; but the 2016 US National Drone Racing Champion-ships are expected to be even bigger.

FPV BASIC EQUIPMENT

Although the following equipment list might look short, there is a large variety on the market, and you should choose you FPV equipment with care. In essence, you will need the following:

- Small, lightweight camera capable of being lifted by your drone
- Video transmitter/receiver matched pair
- Two separate batteries (or adaptive cabling to tap into the drone battery)
- Monitor
- FPV goggles
- Computer

BEFORE YOU START USING FPV

Make sure your basic drone platform is flying perfectly in Stabilized mode before committing to any FPV activity. You should have learned some basic pilot skills; this is important even if you purchased a drone with a camera already attached. As it's a self-build, you will need to check that your FPV system is working before you install it and that your drone has enough surplus lift to carry it. Not all FPV systems are created equal.

In the air we will need a good, stable platform capable of carrying our camera (possibly two), a transmitter to send the video, and a power supply. On the ground we will need a means of receiving the video signal, a screen, and a power supply. If our camera has the capability of being moved (tilt/yaw), then we will need an extra channel on our RC transmitter or GCS that allows the operator to manipulate the camera gimbal.

> **Author's Note:** A *gimbal* is a platform on which the camera sits or that contains a camera. The gimbal has one or more server motors that can tilt the platform up or down as well as perform a 360° circle. Some FPV gimbals have integral cameras all ready for fitting, whereas others are a platform on which you place the camera of your choice. In the hobbyist world, some cameras have been specially developed for PFV work. If you opted to purchase your drone, the chances are you may already have a FPV camera attached.

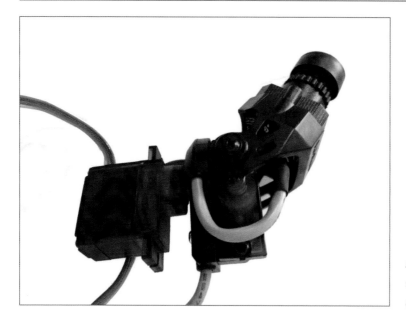

A simple camera gimbal setup providing both pan and tilt.

PLATFORM

The drone carrying your camera is the most important part of any FPV photography; if it fails, your camera is useless. If you have opted to purchase a drone with a camera, then you can be assured that the drone is capable of lifting your FPV system. A 3DR Solo is an unbelievable piece of technology for a great price and can get great outdoor shots at low altitudes using the GoPro camera. You'll need to get an FPV system with excellent functionality and performance while keeping the weight to a minimum. The lighter your payload, the longer your battery will last, and the longer flight time you'll have. Having built a drone, it is for you to decide what PFV system you want; our drone build will carry a system that weighs up to 3 ounces (80 grams), powered from the drone battery.

CAMERA

The cameras used for FPV come in all sizes, and you can purchase a simple, inexpensive CMOS camera for around $45 (£30). On the other hand, if you want quality pictures, you are going to have to spend several thousand dollars. To start, you are better off using a simple system, but if you venture into more serious FPV work, then I recommend a 4K camera whenever possible; that'll give you the highest quality footage and the most flexibility in post-production. The next sections discuss a range of cameras suitable for FPV, from the cheapest to the more professional.

FatShark

I include the FatShark 700TVL CMOS Fixed Mount Camera because it is the one we will be using for the drone we have built, and also later on when we improve the body. It's a great little camera, though a little dated. It is super light in weight, consumes little power, yet provides a great picture. At just 15 grams including the cable that connects directly to your video transmitter, it is small and light enough for most drones.

FatShark 700TVL CMOS camera provides excellent short range video.

GoPro

The GoPro has been around for several years in the outdoor market, but has more recently been used for FPV. Why? Because it lends itself perfectly to FPV flying: it is light, extremely good quality, and cheap. This combination has seen the GoPro fitted to a large number of both commercial and hobbyist drones such as the 3DR Iris and Solo, DJI Phantom 2, and the Walkera QR X350 to name a few. Unlike standard FPV systems, which need a monitor or goggles, the GoPro delivers a live HD video feed straight to your mobile device. You can get a 720p HD live stream from half a mile away (800 m), which is far enough for most FPV fliers. GoPro HERO3s and models upwards all offer simultaneous photo and video capture, so you can shoot high-res stills while you're shooting video.

GoPro, a brilliant camera that is ideal for drones and is used around the world to take extremely high quality images and video.

Professional Camera

The UXG-350 is a two-axis, high-speed, gyrostabilized gimbal with a 360° continuous pan rotation and a 105° to 105° tilt. It contains a day camera with low light capability; as the scene/light fades, an infrared filter is automatically replaced with a clear filter, and the camera switches to black-and-white mode, allowing for operation at a minimum illumination of 1.0 lx. It also houses a thermal camera.

The day camera specs:
- Image device: ⅓-type CMOS
- Effective pixels: 2,000,000 pixels
- 10x optical zoom;12x digital zoom
- Focus system: full auto
- Trigger, manual, infinity, near limit setting
- HFOV (horizontal field of view): 50°–5.4°
- Video out analog: NTSC or PAL and Full HD
- Output: Y/Pb/Pr (optional, ask for quote)

UXG-350.

Thermal camera specs:
- Spectral range: 8–14 um
- Resolution: 384×288
- Digital zoom: 2x/4x digital zoom function
- Operating temperature range: 5°F to 122° F (-15 °C to 50 °C)
- Humidity: < 95 percent
- Video output: 60 Hz NTSC or 50 Hz PAL

Mechanical specs:
- Size: 170 x 216 mm
- Weight: 3½ lbs. (1,600 g)
- Control interface: RS-232, RS485, CAN, PWM
- Power specs: 12VDC, 25W (max)

Video Transmitter and Receiver

As with the camera, the video transmitter and receiver you purchase will have a wide variation. The thing to understand is frequency and antenna, as these two factors can really make a difference to the range and quality of your video stream. The camera sends the signal to the transmitter that in turn radiates the signal out of the antenna. The transmit antenna is an omnidirectional antenna that sends the signal out in all directions. Frequency makes big difference also; the lower the frequency, the better the signal. For example, 900 MHz is best behind trees or buildings as it can penetrate through objects. There are basically four different frequency bands used in FPV analog video transmission:

- 900 MHz
- 1.2 and 1.3 GHz
- 2.3 and 2.4 GHz
- 5.8 GHz

One of the problems with selecting a video frequency is that most likely your transmitter is working on 2.4 GHz, so that is out. The golden rule is not to transmit any two devices on the same frequency—or even close for that matter. GPS operates between 1.5 and 1.3 GHz, so your GPS has to be blocked, which leaves you 900 MHz and 5.8 GHz. The 900 MHz frequency requires a larger antenna, and the quality is not as good (added to which our mobile phone works on 900 MHz in the UK and in some countries the frequency requires a radio operator's license to use it legally). One other thing you should check is your power output, as again this can be regulated. So it's down to 5.8 GHz and, to be honest, whether you're just starting out or are a good FPV flyer, at least you have 2.4 GHz clear for your RC radio system, and you should have no trouble with your GPS.

If there is any issue with 5.8 GHz, it's the lack of penetration, so for the most part you will be flying LOS if you want good reception. You can improve your reception in a number of ways, and the antennae for 5.8 GHz are quite light and small.

We next need to look at the power output of our video transmitter; in general, you are advised not to exceed 600 mW maximum. With the right directional antennae, you will be able to get more than enough range. Anything higher is likely to start interfering with your other frequencies such as your RC and GPS.

Author's Note: No matter what type of video transmitter/receiver antennae you use, *never* turn on the transmitter/receiver without first attaching the antennae because you would be in real danger of doing a lot of damaging your transmitter/receiver. The RF energy needs to radiate or it will burn out.

Video Transmitters and Receivers

When FPV flying started, it provided an experience of poor clarity, with low resolution and a lot of noise from interference. This meant that the pilot's live video view was far from perfect. Analog video transmitters and receivers are still widely used, especially when using a light, agile drone such as those used in drone racing. Added to that, most analog systems are relatively cheap. However, all that is now changing as HD transmitters are becoming lighter and more affordable. Although they are designed for the top end of the market, mainly for those interested in professional aerial photography, the range and quality is far superior to any analog system.

Immersion 600 mW 5.8 GHz A/V Tx

This is one of the most popular A/V transmitters in the FPV hobby world, because it is efficient, lightweight, and really packs quite a punch. This A/V transmitter provides a clean and powerful transmission that is compatible with all FatShark and Immersion RC products. The unit includes a built-in super quiet switching regulator, which powers not only the transmitter, but a connected 5v camera. Added to which you can power the transmitter and camera directly from your flight battery. We will be using this transmitter in the FPV setup for our drone. It's a basic setup and a great place to start.

Since I am using an Immersion transmitter, I am going to use one of its receivers—in this case a Duo5800. This receiver provides clean video in even the harshest of RF conditions, and many 5.8 GHz FPV frequencies are also fully supported, including Boscam, Team Black Sheep, DJI, and others. The dual independent receivers greatly increase usable range, especially when using two patch antennas to cover a wider flight area.

A simple Immersion 600 mW 5.8 GHz A/V Tx setup.

HD Wireless Link

DJI introduced the Lightbridge for its Phantom range, which really improved things; 3DR has recently done the same with its launch of Solo. Nevertheless, the unit I favor is the CONNEX, mainly because it's not tied to one brand of drone and works within the 5.8 GHz frequency, so it does not interfere with the RC transmitter. One drawback is the weight, which is some 4½ ounces (130 g) for the air unit, and the price: $1,599 (£1,299).

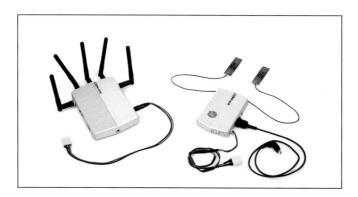

The CONNEX HD Wireless Link is plug-and-fly out of the box and perfect for aerial video production and cinematography. With a range of up to 3,300 feet (1,000 m) line of sight, it has zero latency and offers real-time encrypted video. It also provides a MAVlink-based telemetry OSD view.

FPV with Mission Planner

There is nothing to stop you connecting your video receiver to Mission Planner on your laptop. You simply need to purchase a video grabber to convert the signal and connect it to a USB port on your PC or laptop. Although this does tend to cause a lot of cables hanging around, you can always build the whole unit into a single entity (see

Video downlink video can also be displayed in the Mission Planner HUD.

Chapter 9). This works extremely well for general FPV flying, as you have all the visuals and control open to you in one place. You can even connect a joystick to the laptop and develop your own full GCS not too dissimilar to those used by the military.

Pilots View

As I previously mentioned, I feel that you are better off flying FPV when you begin using a monitor rather than goggles, but again that's just my personal preference. There are several excellent screens on the market now designed for FPV, many of which have a built-in 5.8 GHz receiver. My personal favorite is the Black Pearl. As for FPV goggles, you will constantly hear the words "FatShark" being banded around. I believe the company was the first to introduce goggles for FPV way back in 2006. Although I am not a big fan of goggles, I do have two sets of FatShark that I occasionally use. The goggles themselves are similar in function to the monitor as they incorporate a receiver.

Black Pearl monitor with built-in 5.8 GHz receiver is an ideal choice when you start to learn FPV flying.

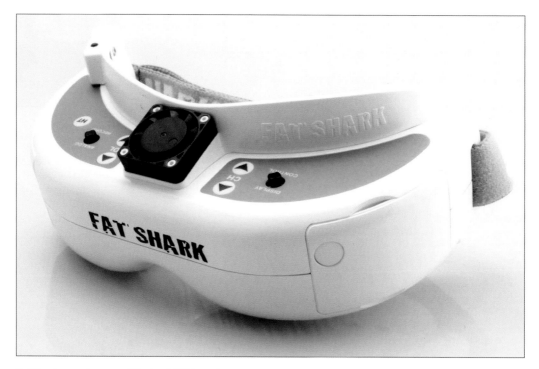

FatShark remains one of the best FPV goggles.

Antenna

I am no electronic specialist in antenna theory, but during my military days I used to send Morse code from a small shoebox transmitter deep in the jungles of Asia directly to Cyprus and then on to the UK. I soon learned that getting connected came down to the half wave dipole aerial being large enough. The same problem of range confronted me again when I started flying FPV. What you get is not always what you want. For example, when you buy your transmitter or receiver, it will most likely come with one or more antennae. In my experience, they are not very good, and you will need to improve them to get half decent range and picture quality.

The simple linear antennae radiate outwards from the pole in a kind of flat, doughnut shape; this radiation bounces off anything it hits and causes multi frequency interference. These antennae look like the ones you find on your wireless router at home. You will also see the words *directional and omni-directional* when searching for your perfect antenna. Directional antennae transmit and listen in a specific direction, whereas omni-directional antennae transmit and listen in all directions. I strongly recommend using circular polarized FPV antennae.

One other way of improving the video range and quality is to install a booster system between your antenna and the receiver. I recently purchased the ClearBoost from a Kickstarter program which claims to enhance your FPV/UAV/drone video reception. I have yet to try it, but published results sound encouraging. Celestial Reach, which produced the device, says the dropouts in signal quality are greatly reduced. Hopefully this will help those who fly in drone racing.

> **Author's Note:** No matter what system you are using for your FPV, in many instances you can increase range and quality by simply putting your receiver higher. I use a basic metal spike that I press firmly into the ground (tripods tend to get knocked over by people or wind), leaving at least two feet (a little more than half a meter) exposed. I slide on a neat-fitting adjustable painter's extension pole, on top of which is my receiver. The extra 10 feet (3 m) in height makes a big difference to your LOS.

Antenna Tracking

As with satellite TV, the dish has to be aligned with a geostationary satellite in order to receive a good signal. Because our drone is moving, we need to be able to follow it. Antenna tracking simply means tracking the position of your drone as it flies, and in doing so you receive a much better control and video connection. In order to do this, you are going to need some extra equipment that involves pan and tilt antennae moved by servos. The overall weight of the type of antennas used, plus the added components such as frame, battery, and telemetry unit, will determine the type of servos required. Although it is possible to build your own setup, I would recommend that you purchase a ready-made or kit form antenna tracking system.

Some ready-built systems offer a 360° pan with a 90° tilt. If you have never done this before, you will need to follow the setup procedure closely. Once setup, you will need to make sure that any movement of the unit does not involve snagging any wiring. The antenna tracker is simply following the location of the GPS unit in your drone. There are various ways of tracking your drone, including using Mission Planner.

FIRST FLIGHT USING FPV

I am a great believer in using a monitor system for your first flight, as flying with a monitor is not as frightening as flying using FPV goggles. This is not the case if you are drone racing, in which case you will almost certainly be using goggles. You also need to get used to your environment, so choose a place you know well, such as your local flying field, large back yard, or farm field: any open space that has no dangers to others or to your drone.

Before you start flying, familiarize yourself with things such as trees and buildings close to your flying space. Gauge the height of everything and its relation to the direction in which you are standing. Try to capture this image and keep it in your mind. In particular, pay attention to any linear features such as hedge rows, fencing, streams, or anything that is significant. Noting these small landmarks can help aid your FPV flight as you will be aware of the direction in which they are running and their height perspective; in short, you will have a reference point. Early pilots in WWI used similar methods to navigate, noting a list of prominent features to guide them. Make sure you keep the drone flying "front away" from you so you and the drone are looking in the same direction. Orientation is one of the biggest challenges for new pilots and is even harder when flying FPV; things get confusing very quickly.

FPV requires a lot of patience, so don't rush it. Start off small, keeping your drone under control rather than worrying about FPV. Have it taking off and hovering at around 16 feet (5 m) before flying about 100 feet (30 m) out toward a small marker or bush. Then rotate the drone so you can see yourself and fly back. At this stage you want to concentrate on estimating your height, direction, and speed. Stay alert to your flying and if in doubt switch to Loiter, remove your goggles, and concentrate on flying your drone safely home.

At each new location, face the direction of flight and look at what is in front of you. Then cast your eyes both left and right until you *know* what you will be seeing when you put your goggles on. Lift off and hover your drone about 16 feet (5 m) high and then slowly turn to the left and right to see if you can observe the same features you did without the goggles. Situational awareness and orientation are key to good FPV flying. Having a camera gimbal with the ability to pan and tilt or placing more than one camera on the drone can help, but in the initial stages they can also detract from your flying.

You have to remember what is carrying your camera; it's a drone, which has spinning and very noisy motors, and it most likely has great big legs for takeoff and landing. Your propellers and legs can cast shadows on your video footage or do other strange things to your video—be aware of the angle of the sun. Beware when you take off from a dusty place as the propeller wash will raise a lot of dust; always check and clean your lens before takeoff.

SLOW, SMOOTH, AND SHORT

Commercial filming requires you to avoid large melodramatic or fast moves, as these lead to shaky footage and possible crashes. Keep it slow and smooth with short, steady, planned shots of no more than 10 seconds, but start shooting well before your objective and stop well after it so you give the editor some room to maneuver. If the sequence requires a longer time, do several runs that can be well edited. Do not try to keep the camera in one spot; ever so slowly keep it moving in one smooth direction—tiny and gentle stick movements produce by far the best results. Before you go for the real thing, practice and rehearse based on a simple plan; save your video and see how you can improve on it when you shoot the real thing.

When flying FPV you will usually have a specific aim in mind—for example, to video an event. The first thing to establish is a direct LOS. It's always easier to fly directly out and back again. Check that there is nothing in your LOS that may cause an obstruction; that means you can fly to your objective, film it, and fly back still filming if required. Establish how you are going to set up and view your video (or snapshot photo) footage. If you are going to use goggles, then make sure you have a spotter (a person who can watch the drone) with you just in case some hazard appears you are not aware of.

Don't be afraid to cancel a shoot. Wind is not your friend. If it's too windy outside, delay the shoot. Even if your drone survives a high wind environment, your footage probably won't. We are not testing the drone's capabilities, we are looking for the perfect shot. Practice taking videos before you commit yourself to any professional work. Learn to master aerial camera pans.

Bird's Eye

Have your camera pointing straight down on takeoff and slowly climb. You can experiment with slow circular turns of the drone when it is either stationary or while climbing.

Side-Slide

This is similar to the film industry using a dolly. As the subject moves, so does the camera. You need to maintain altitude while moving in the direction of the target. If your camera has a zoom function, you can try zooming in or out as you slide.

Fly-through and Fly-over

A fly-through may involve going through a small opening; chasing a running person through a forest would be a good example. This requires a lot of skill, and a little bit of drone racing might come in handy for this technique. Fly-overs can produce some really great aerial shots. It simply means sending the drone out where it is obvious it is an aerial shot. Flying at head height over a cliff so that the "drop" effect is clear to see can be spectacular.

Drones can capture incredible images and video at a near perfect height.

STRUCTURE ORBITS

This move can be done automatically, but the best results are done manually. It simply means flying your drone up to the top of a structure and doing a 360° around it. The skill is to make this look smooth while keeping a corresponding distance from the object. You will need to have good yaw and roll skills for this.

SUMMARY

Flying FPV is extremely satisfying and, as with our drone build, it is best to be able to build your own system. That said, the easy route would be to purchase a drone with an integrated camera gimbal, but then what do you do if it goes wrong or crashes? If you intend to use FPV for drone racing, you will most definitely need to know how to repair your drone and your FPV system as you can expect a lot of crashes during your learning curve. You will need to get used to wearing goggles; it's not as easy as one would suppose, and you can suffer from vertigo or motion sickness.

If you're using FPV simply for recreational fun, then you will not encounter too many problems, and the experience you gain will all add to your flying skills. Remember, safety is the key, both for others and the survival of your drone. For those keen on aerial photography, the situation is different, and you will need to start off slow, building up to more advanced shoots. Check out this example from Epic Drone Videos, which shows that it's—simply outstanding what drones can achieve: youtube.com/watch?v=AoPiLg8DZ3A.

Chapter 9

DESIGNING YOUR OWN DRONE

et's face it, the finished product we have built, while fully functional, is not the best-looking drone in the world. Not that it really matters, but it's like having a Ferrari with no body shell: no one will notice you. So, as a final stage, we are going to strip down our drone and make a few modifications, including designing and providing our drone with a beautiful body.

At this stage I must warn you that although designing a drone body can be done for free, it will take you some time; likewise if you 3D print your drone body, it will most probably cost more than building your original drone. Putting all that aside, let us look first at the options.

We could search the Internet for a free quad-copter drone design and use that (with the relevant permissions if applicable). We could sit down and design our own using free CAD drawing software such as Sketchup. You could use the example I have provided with this book. You could also consider purchasing a spare body of an existing commercial quad-copter, as many manufacturers sell these as replacement parts. No matter what path you choose to take, at the end of the day as long as you finish up with a shiny designer body, it will be worth it.

My finished drone design into which I have shoehorned all my components.

Above is the completed new body for my drone, fully assembled and ready for testing. Its design took several stages, and I considered many changes, but each time returned to something I knew would work. However, what looks good in theory does not necessarily work in practice. Folding arms can be made to work, for example, but the locking mechanism must be perfect in order to avoid vibration. Likewise, you need to think about all the components and check to make sure that they will fit in with your design. Finally, there is the battery. It's big and it's heavy, so it must fit centrally in order to preserve balance.

The real problem comes when we try to cram all our electronics into our new body shell, and I warn you now: positioning of components when designing your new drone shell is the most important thing to keep in mind. Every component of your drone works with every other component, and the frame is the one component that holds it all together. So, when designing your frame, there are a few basic rules you need to take into consideration:

- It's a quad-copter, so it must have four arms asymmetrical from the center and far enough away from the body to avoid the propellers touching.
- Make your body shell fairly rigid so as to minimize motor vibration. Take extra care when designing the motor mounts.
- Keep your electronics and the battery in the center so that the drone is well balanced.
- Consider the position of all your components: camera, GPS, flight board, battery, and so on, in order to avoid electrical interference.
- Your battery is a large and heavy part of your drone; make sure you can connect and disconnect it with ease. A battery falling off in flight may cause the drone to crash.
- It is best to design your arms and motor housing so that the wires from the motors and ESCs are hidden.
- If you are using a 3D-printed frame, put in enough internal slots and shelves to receive all your components.
- Finally, what material will you use for your new frame? Reducing weight is vital.

Design for 3D printing.

Start with how long the arms will be, which is controlled by your motor size and the proportionate propeller size. In our case we are using 9½-inch propellers, which means the arms point to point (distance from the center of one motor to the center of another motor) must be at least 10 inches (16 cm) apart and above the body to stop them from touching each other or the frame. Then decide how large you need to make the center body in order to comfortably fit all your components. On our current drone, this is approximately 12½ inches (32 cm), but now we want to try and reduce this to 10 inches (25 cm).

Next, you may want to look at your frame configuration. Do you want to stick with your X configuration or change it for **+,** which will look like the following image? The difference is in how the motors are controlled; in the **+** configuration you have one motor controlling one direction whereas in the X configuration you have two. Likewise, the **+** only provides one motor for extra thrust compared to the X, which has two motors, pushing your speed up. Having flown both, I think I prefer the simple X type configuration for no other reason than it seems the best format.

Before you set off on your journey as a designer, we must make sure we understand what it is we are designing. In essence, we need to make sure that the final weight of our new drone frame, together with its motors, electronics, and battery, will result in it flying nicely using the motor size and propellers we want to use. That is to say, we could design, build, and assemble our beautiful looking drone only to find out that it will barely lift off the ground— let I alone fly.

The material you use for your frame will very much depend on your design. If the design is simple, you can build with off-the-shelf carbon rods and sheets, or aluminum. However, if you choose to design a complicated frame, it will almost certainly have to be 3D printed.

Author's Note: My advice is that you should avoid building with aluminum or MDF, as they are unnecessarily heavy. True, in the past, many people have used these materials with great success, but today they are a little dated for quad-copters.

It might help if I explain what I did when designing the second version of the SQ-4. We had tried many previous versions, which in the end turned out to be far too complicated. The main problem was that we tried to "jam pack" too much new technology into a small body and, although this worked in theory, in reality the drone was just too futuristic to fly smoothly because too much was happening at the same time. Another main problem was the communications—we opted for Wi-Fi and with this came all the problems of frequency saturation and unreliability.

Then about two years ago I went back to the drawing board and redesigned the whole system of the SQ-4, starting with something not unlike what we have just built, albeit more advanced. The first bodies were made using a 3D printer at our local university.

This is the model I designed and I had to make four changes to the design because of faults, making the end product very expensive. The faults were simple

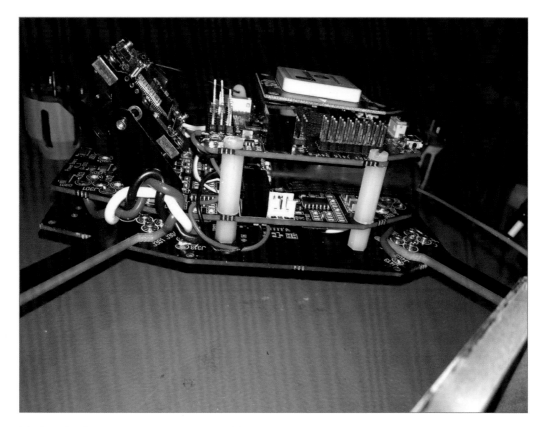

SQ-4 version 2 was an extremely complicated drone to design and construct, and it lacked the ability to fly any great distance. There was no way a non-expert could repair this drone if it went wrong. Additionally, the flight range was limited.

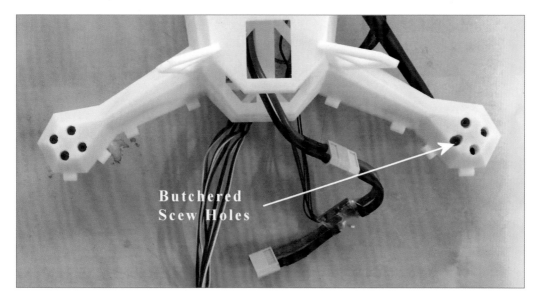

Butchered screw holes due to a silly mistake meant we had to print a new body—a costly error.

things that we got wrong, such as designing one arm and repeating the same design for the other three. It was only when we came to fit the motors that we discovered that two of the arms had the screw holes in the wrong place on the motor mounts, and "butchering" the mount was not an option. This simple mistake meant that we had to correct the CAD and reprint.

You can see in the picture at the bottom of the previous page, the center internal plate on top of which we placed our autopilot. The two smaller plates at the back were for the video transmitter and the telemetry unit. Toward the front you can see the small block with two screw holes where we placed the battery connector. This allowed the battery to be pushed in and out with ease. The hole is for the down-looking landing camera, and the raised section on the top cover is for the GPS housing, which had a shielding over it to reduce interference from the autopilot. The arms were hollow and spacious enough to take the ESCs.

On the assembled model you can see the 2.4 GHz RC receiver on the top plate and the Safety button just behind it. To the front we have the LIDAR, which we adapted for sense-and-avoid using an advanced algorithm while connecting it directly to the main processor instead of the I2C. Below that are the two cameras, one at 45° and the other at 180°. The model flew extremely well, and the only trouble we encountered was the breaking of the 3D body when it had a sharp knock. We remedied this by 3D printing in a stronger and more flexible material.

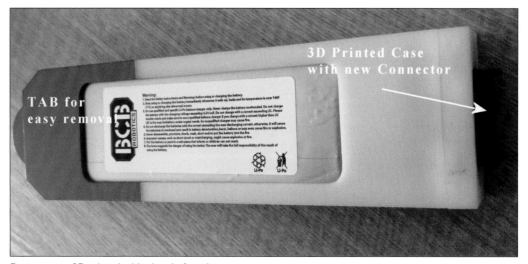

Battery case 3D printed with plug-in female connector.

The next part of our design came with the battery housing and charger. I designed a body for a battery size and power that fit the criteria of the drone design. That is to say, it needed to power and fly my drone for the desired duration of 25 to 30 minutes. I incorporated the female connector to marry with the male counterpart in the drone, making sure the connector I chose would be strong enough to hold the battery in place while in flight. I also added a simple spring tab to the drone's body, which locked the battery in place. The final design added an extension to the battery casing to aid removal: a battery change time of less than 5 seconds.

The battery charger case was also 3D printed and housed a standard charger fitted to a male connector. Hard foam guides helped the battery slide neatly into place for charging. The labels were designed and printed on glossy sticky-backed paper using my computer and printer.

The battery charger was designed to marry with the battery housing, making charging simple.

Author's Note: Some parts lend themselves well to 3D printing, such as the battery casing and charger unit. All of the units I made eighteen months ago are still intact and functional.

I obtained the GCS from UAVision in Portugal and made a few modifications so that it would communicate with the drone. The UXC-850 GCS is a small, hand-held GCS that incorporates FPV and a lot of redundancy; that is to say, if one system fails, the other remains active as both work in tandem or individually. All major modes including Arm, Stabilize, AltHold, Loiter, Auto, and RTL can be performed either by the RC or via the DroidPlanner app. The camera gimbal is controlled by two joysticks under the unit using the middle and third finger. A gimbal is a device that permits the drone to incline freely in any direction or suspend it so that it will remain level when its support is tipped. This makes for good flight control and very smooth camera gimbal operation. There are also easy-to-reach buttons that help to operate the lights and switching between cameras. LIDAR is operational when the drone is in AltHold mode and is used for sense-and-avoid.

Although bulky compared to most commercial RC controllers, the UXC-850 has both RC controls and telemetry working together, offering redundancy if one system fails. The unit fits nicely in the hands and is simple to operate in both manual and Automatic mode. The folding video screen provides excellent FPV from either camera, while the mapping and flight plan can be followed on the mapping screen. Its main features include the following:

- Two sets of joysticks for simultaneous UAV and gimbal control
- Access to the UAV telemetry parameters
- Telemetry link compatible with MAVlink (433 MHz, 868 MHz, or 915 MHz)
- Control Link: 2.4 GHz, 12 channels PPM or TTL Series
- Video link: 5.8 GHz (boosted)
- Access the operational site map with UAV location and route
- Possibility of using UAVision software or open source software
- Display: high-quality 7-inch TFT LCD (1200 x 1920 pixels) HDMI output
- Capacitive touchscreen Multitouch, 16M colors
- Power and battery: 7 hours
- Operating system: Android 4.4 KitKat
- Battery: 1 unit rechargeable LiPo 7.4V/8000mAh
- Size: (opened) 322 mm x 259.2 mm x 189.4 mm (closed) 322 mm x 192.9 mm x 111.2 mm
- Weight: 4⅓ pounds (1,970 g)
- Resistance: IP 64

3D PRINTED

There are many companies that will offer to print out your CAD drawing, and I advise you to shop around before making your decision. There are several places that can help, such as your local university, as most engineering departments will have an excellent 3D printer, if not several. If you go down this route, you may want to join a university activity that allows you to use the 3D printer for free (such as taking an evening class in CAD design, for example). If you go direct, they will assume you are a commercial outfit and will be charged accordingly. You might try a local firm that has a 3D printer or go to a dedicated 3D printing service, both of which will be expensive. Finally, you might consider purchasing your own 3D printer, as they now appear in places such as Media Markt (Europe) and Best Buy (US). It seems there's a 3D printer for everyone on the market today. However, it can be a bit confusing having to choose between all of the printers and what systems they use.

Author's Note: Personally, I would stick to finding a helpful university, as staff there may be able to advise you on the best material for your 3D printing. You will also find that many manufacturing companies have a lot of redundant 3D printers sitting idly. The pace of advancement within the 3D printer world has been similar to that of drones, and last year's model is out of date. Two of the companies I associate with have $30,000 3D printing machines doing nothing because they have just purchased a new one for $5,000 that is 100 percent more efficient.

CARBON FIBER

Molding your new frame in carbon fiber is no cheaper than going the 3D printer route. First you need to have a mold made, which in turn will produce the relevant parts for you to put together. This is a skilled job that requires professional knowledge. Even when your parts are ready, you still need to assemble the carbon fiber and fit any internal plates with resin/glue. The main advantage of carbon fiber over 3D printing is one of strength, but beware: a strong knock can easily break any joints in the carbon fiber.

The main advantage of 3D printing over carbon fiber is that you can design in all your internal parts and screw joints perfectly, so that it is all ready to receive all your components. The downfall of both processes is that making an error in your design means you will have to butcher your finished model or re-design the error out of the equation and reprint or make a new mold.

REMODELING OUR DRONE

If you want to proceed and remodel the drone or build it into something that looks better, you are going to have to make some modifications other than just

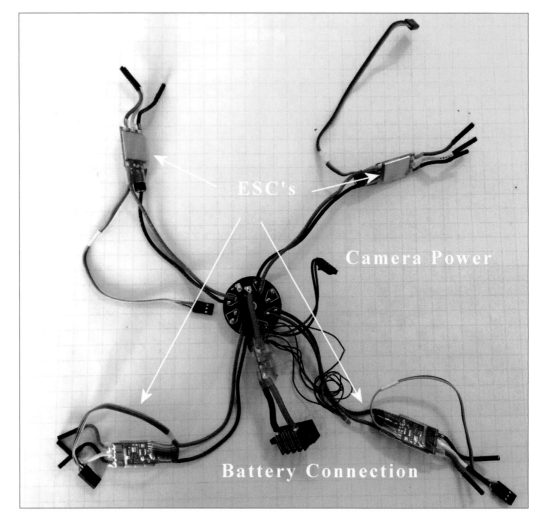

Here the picture shows the four new ESCs soldered in place on the power hub. I have also attached the battery connector leads plus a power lead for my camera which I intend to add in order to create a FPV drone. Finally, I cut down the 3DR power cable and attach this to the power hub—the other end is connected to the male battery connector that will fit inside the new frame ready to receive the battery.

designing a new body. The current F450 frame base and top are much too large to fit into the new body, which means we need to reconfigure the whole basic layout. The major components, such as the motors, propellers, autopilot, GPS, RC receiver, and so on, will be reused, but we will need to change the ESCs and the bottom board, which means more soldering.

When you purchased your tuned propulsion system, you will have found in the box a power hub, which, if you look closely, has a similar layout to the 450 bottom board onto which we soldered our ESCs and power cables—albeit the power hub is a lot smaller.

The first thing we need to do is to purchase four smaller and lighter ESCs, mainly so they will fit better inside the arms of our new body. I chose the Castle

Creation 25 QuadPack, because these are not only smaller but are ⅓ ounce (9 g) lighter at 0.6 ounce (17.8 g) each, saving me 1¼ ounces (36 g). You will note that the cables on the new ESCs are also thinner, and the bullet connectors are smaller; therefore I will change the bullet connectors on the motors from 3.5 mm to 2 mm to match—again saving several ounces.

With all our parts together, I am now going to solder them. You will notice that the power hub has seven-point sets that can be soldered, plus one set to connect power.

Carefully slide the power hub into your new body; for me, this means sliding it between the battery and the autopilot plate. Next, push the individual ESC cables up each arm until they appear at the motor mount. There should be enough slack in your cable so you can connect to the motors without too much trouble. I have added three double-sided sticky mounts to the underside of the power hub to hold it in place.

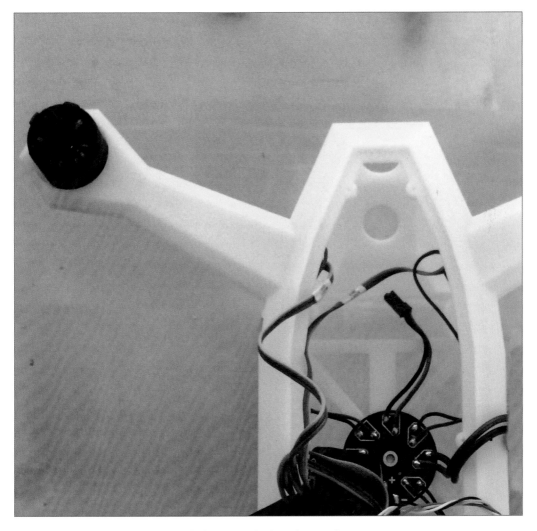

Place the completed power hub and ESCs into the body and secure it.

You can now fit the autopilot; I have replaced the Pixhawk with a stripped down APM 2.6 as the pins face upwards making connections easier.

The new body is finished, recalibrated, and on its first flight.It flew extremely well and required very little tuning. Note the nice coat of paint that makes it look extra special.

You will now need to strip all the motors and components off the 450 frame that we previously built and transfer these to the new body. Connect the ESCs and screw in place, maintaining the same order.

My design has a small notch that will accommodate the male battery connector, which is held in place by two 8 mm screws. This should be fitted securely, as it will experience a lot of wear as batteries are put in and removed.

Next, I am going to fit the camera through the hole created in the front. To secure this, I add a Velcro patch which fits around the front of the camera and sticks to the inner frame, holding it firmly in place while providing some anti-vibration. Again, during this design I created a slot behind the camera where I placed the video transmitter. The transmitter sits upside down with the antenna screw exposed through the bottom of the frame and secured in place with a nut. Connect the camera cable to the transmitter, and also the extra power lead we added for the camera system.

Place the telemetry unit and RC receiver on the two plates at the rear of the body; again, use Velcro to hold them in place. The antenna for the telemetry will fit through the hole at the back while the RC receiver will go up through the top cover when it's fitted.

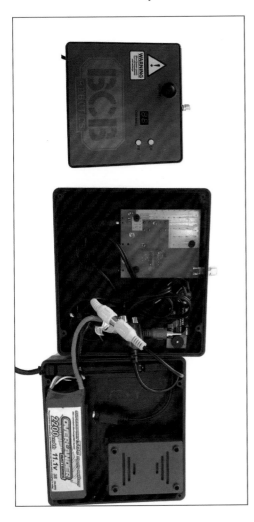

We are now ready to fit the autopilot, having removed it from the 450 frame. I suggest you use new anti-vibration pads to ensure you get a secure fit. Place the autopilot as central as possible on the plates provided and press down to secure.

Next, we will need to connect our ESC cables—again in the same order as on the original. Be careful to tuck as many of

The box contains a stripped-down video receiver that is connected to a video grabber and is powered by a battery with its own charger. I added an on/off button for when it's not in use and a small hole to enable me to charge the battery. Normally I get around three to four hours use out of the system. I plan to attach this to my painter's pole so I can elevate it and improve range and quality; for this I will need a USB extension cable. Note the maximum recommended length for a USB extension is 16 feet (5 m).

the wires to the outer side of the body and push the ESCs up into the arms until they "jam" fit.

You can now connect your RC receiver and the telemetry to the appropriate connections on the autopilot as mentioned in Chapter 4.

The top cover of my new drone (which is removable) contains the GPS. The design has a specific raised spot that the GPS (case removed) slots into. You will need to shield this with some sticky-sided tape such as EMI copper foil, which seems to work for me. Leave the two sets of wires exposed so you can fit them to the autopilot just before you secure the top cover onto the body.

And that's just about it! You can, of course, make some modifications that will make life a lot easier, but that's entirely up to you. The main one would be to fit a USB extension to the side of the body so you do not have to open the drone each time you want access to the autopilot. External USBs make life a lot easier when it comes to configuration and calibration. You can now connect your new drone to Mission Planner and carry out a complete setup using the Wizard once more, as noted in Chapter 5.

FPV SETUP

I have never been a fan of wires, and the less I see of them the better. Unfortunately, the world of self-built drones is inundated with wires. So, as far as I can, if I am able to shorten the length to hide them, I will. Although the camera in my new drone will happily transmit directly to a monitor or a set of FPV goggles, I want to set it so I can view the video in Mission Planner.

This will involve using a video receiver connected to a battery and a video grabber which in turn is connected to my laptop—a lot of cables. So I decided to build my own receiver set, which I could simply connect to my laptop.

SUMMARY

While flying remains the primary motivation for most drone enthusiasts, there is a growing number that truly love designing and making their own drones. The one great thing is that you do not need to consider any aerodynamics in your design; true, you will need symmetry and balance, but that's about it. Weight is not a factor if the propulsion system and power supply are equally provided for.

You will need to understand a little about 3D printing, especially what materials are available and best suited to your design. You will also need to know the limits of the process you choose to use. 3D printing is on a similar level with drones; it is expanding its capabilities literally by the day.

Chapter 10

THE FUTURE OF DRONES

Now that we have built our drone, and possibly remodeled the body as I did, there is a lot more we can think about. The drone world is expanding at a phenomenal rate, and 2016 may well prove to be the "Year of the Drone." Yet within this book we have barely scratched the surface on so many uses and abilities where drones can be deployed. As technology improves, so will drone functionality, and they will further become a part of our world. Drones will fly safer, in swarms, and maybe one day from continent to continent. This book is only a glimpse into the world of quad-copters, and we must not forget that the word "drone" also applies to fixed-wing aircraft as well. While this book is primarily about quad-copter drones, I would remiss not to mention the fixed-wing aircrafts under the umbrella of drones. There is a future for both, and as we progress we will see the value that technology can offer in the realms of safety, sensors, and communications. This chapter offers a short insight to some of these developments.

SAFETY

Many of the new functions we will see in the future will entail some form of safety; not just in where and how we fly but within the drone's capabilities itself. When you launch Mission Planner, you will see that just about every major airport on the planet is highlighted within a circle as a warning. This also appears in other GCS software and is there to provide a safety net—do *not* fly in these areas (without authority and permission from air-traffic control). This coupled with further drone safety instructions issued by some countries all helps to prevent accidents.

Ways of limiting the damage when drones do fail will also be a major feature; this may include adding an automatic parachute or simply fastening the drone to the ground with a tether. Adding a parachute to your drone which, in the event of malfunction, will deploy and bring your drone safely to earth makes sense. Fitting sensors such as sonar or LIDAR that will keep your drone at an even height above ground or help with sense-and-avoid will also improve drone safety. Using newer communication links that allow one drone to be sent on a mission by one operator and received by another several miles away is also a possibility. As previously mentioned, the rate at which drone technology is advancing is dumbfounding, and it seems that every week someone comes up with something new or makes a major improvement.

PARACHUTE RECOVERY

A growing number of parachute systems are becoming available that will automatically deploy if your drone gets into trouble. This applies to both fixed-wing aircrafts as well as multi-copters. Parachutes are placed so they can deploy without any risk of entanglement and, in many cases, involve the drone being turned upside down and away from the propellers. Another consideration is the speed of the deployment, which in an emergency should be within milliseconds. Finally, the parachute should be large enough (considering weight is a major issue when flying) to safely land the crippled drone.

There is no doubt about it: the use of parachutes on larger and heavier photographic drones will help prevent a lot of casualties in the event of an accident—especially when used over sporting events. A recent news item highlighted a drone that was filming at a skiing event in Austria; it fell to earth and crashed just inches away from the Olympic Silver medalist, Marcel Hircher.

Parachute recovery on the Hubsan X4. This top-end quad-copter from Hubsan not only sports some excellent FPV features for photography, but it also has a fixed parachute. If there is a collision or accident, the parachute is automatically deployed. The great thing is it can be reused over and over.

TETHERED DRONES

The idea of tying aerial observation platforms to the ground isn't new; both Britain and Germany used them in WWI. The US still uses tethered balloons for radar along its borders, and the military recently launched the first of two 220-foot long (67 m) tethered airships that will float up to 10,000 feet (3,050 m) over Maryland. They can stay aloft for up to 30 days and can track cruise missiles up to 340 miles (550 kilometers) away.

The main advantages of a drone tethered to the ground are safety and duration. That is to say, it will not fly away or, if used properly, harm any other aircraft or person. As for duration, in theory it could say up for days doing whatever task it has been programmed with.

Basically, tethering means putting a drone on a long cable which uploads its power supply and downloads its data. Naturally, there is a little more to it than that. For example, additional ground support is required. The drone needs to be large enough to lift the weight of the cable that is powering it. The cable needs to be of sufficient strength so as not to break. Entanglement problems and weather conditions need to be considered so that the drone does not fall out of the sky. If for any reason the umbilical cable breaks, there is a need to make the drone instantly land on a backup battery system. Yet these are fairly minor problems to overcome.

Advantages and Disadvantages of Tethered Drones:
- Duration
- Safety
- Rapid data transfer
- Dedicated safe flying area
- Better response to adverse weather
- Does not require GPS
- Restricted mobility
- Tether adds complexity and could get tangled
- More infrastructure needed on the ground

Given that a tethered drone could be positioned to take off from a pre-defined area—one that has FAA approval and one that manned aircraft know about—there are few limits to the possibilities. Height is only restricted by the cable, camera system, and prevailing weather conditions.

Aerial surveillance is an exceptional tool: it can aid firefighters, watch and control large civic events, and help in search and rescue. As many incidents that require aerial surveillance actually cover a small area, why would we risk a free-flying drone over that of a tethered one? Location security is ideal for a tethered drone; a drone sitting at 400 feet (120 m) and fitted with an array of both powerful day and night cameras could do the work of ten security guards (visually speaking).

It's not just surveillance; tethered drones can act as visual warnings with flashing lights; they could be used as rebroadcast stations, improving communications dramatically. Agriculture could benefit significantly, as crop cultivation could be monitored without the farmer leaving his home. Finally, the media community could certainly take advantage of a tethered drone at fixed sporting events such as racing, football, or any other stadium event. Once airborne and stabilized at around 300 feet (90 m), the drone could broadcast quality HD video with a 360° live aerial view being transmitted through the tether directly into the broadcast system.

Tethered drones can be thought of as a grey area. The FAA doesn't consider tethered drones—which can be controlled from the ground—to be any different than free-flying unmanned aircraft—meaning they must submit to the same strict standards. However, I think I am correct in saying if you simply tie your drone to a line (with no control) then you are not required to be covered by FAA rules. It's a little bit like the rules in the UK: fly for fun is fine, fly for commercial use and you need a Certificate of Authorization. I believe the FAA takes the same stance with tethered drones (but rules tend to change, so check with the FAA at faa.gov to be sure). If tethered drones are to become a part of the drone community and used without FAA authorization, then several things need to be addressed.
- Guaranteed safe tether
- Maximum tether height

- High wind sensor
- Automatic tangle sensor with rapid retract
- Maximum takeoff weight
- Auto-land in the event of flyaway
- Special tether pilot training

It would be possible to define and program a tethered drone to launch automatically to a pre-set height from its ground control unit. A simple control of the tether cable length and an automated startup procedure would annul the need for much training. The drone, having reached the desired height, could be instructed from the ground as far as camera direction and zoom are concerned (according to the user's requirements). Adjustable height would be achieved by simply letting out or reeling in the cable. Automatic safety could be imposed if the weather caused the drone to drift too much, as could any onboard problems with the electronics. In the event of a cable break, a small onboard battery would automatically land the drone; likewise if there were a cable entanglement, the crash and possible third-party damage could be reduced by the cable reeling in at raid speed (that is, a faster than natural free-fall). A safety net coupled with a cable motor stop would minimize damage to the drone itself.

Author's Note: I am presently involved in working on a tethered drone for the British Police. The drone itself is some 15½ pounds (7 kilograms) in weight and carries both day and night cameras fitted with a 36x digital zoom. The ground unit will supply the power to keep the drone airborne for up to 24 hours, with the drone itself having sufficient onboard power to land should it become detached. The ground base will have data outputs to our UX400 GCS, plus input for generator power. The cable is military grade released or retracted, with a variable speed motor. The motor is capable of retracting the drone at high speed, dramatically reducing the area in which it can fall should it have a motor problem. One other safety option we are looking at is an automatic parachute release on the drone in the event of a cable break and flyaway. The whole unit, including the drone, will fit into the luggage space of a standard car.

LONG-RANGE COMMUNICATIONS

Imagine controlling a drone in flight just by holding your iPhone out in front of you. You tilt it one way and then another or point on a map and make the drone fly there. Well, it's been done, and there are some cheap commercial models that will do it—but flight is limited to a few feet. In 2014, while at the USSOCOM exhibition in Tampa, our 1-pound (500 gram) SQ-4 drone was controlled by another

These drones were made for MIT who with some alteration used them for drone swarming during the 2015 Drones for Good Challenge in Dubai.

member of our team at a similar exhibition in Washington. Likewise, students from MIT and Boeing jointly developed an app that was successful in controlling a small drone flying around a football field some 2,500 miles (4,025 kilometers) away. In the UK, engineers from the University of Southampton built a near silent drone that can be assembled by hand in minutes without the need for tools. Known as SULSA, the drone was created (using an air-frame) entirely by a 3D printer and weighed just 6½ pounds (3 kilograms) with a wingspan of 5 feet (1½ m). The British Navy allowed it to be launched from a ship after which it successfully made its way to the shore.

DRONE SWARMING

I must admit that one of the greatest fascinations for me is that of flying more than one drone at a time—in fact, flying a swarm of drones. I know many have tried this with various results. Mission Planner supports limited "swarming," that is to say, the amount of drones you use, but I see no reason why numbers should affect the possibility of formation flying. My theory has always been similar to others, control one drone (the prime) and let it disseminate the instructions to the other drones.

For example, if we have a cube formation algorithm that places each drone within its own box [allowing for X, Y, and Z movement of say 65½ feet (20 m)], then the number and formation of boxes within the cube could be controlled by a single drone: the prime. In an ideal world, the prime could be changed at any time and reassigned to any drone upon whose command the rest would follow. The cube formation could be linear, staggered, or arrowhead depending on the task. This theory is used to some degree in Mission Planner when the prime is flying in manual or Auto while the rest are flying in Guided mode. The position of the prime is relayed to the other drones, which adjust their position accordingly. However, Mission Planner does have some swarming instruction; QGroundControl and APM Planner 2.0 are both built on multiple-vehicle architecture.

SENSORS

Sensors come in a wide range of types and cover all kinds of aspects, from those required by the military to those required by farmers. That is to say, a drone can carry sensors whereby the camera can detect and individual and lock on top them—the military use such devices. A farmer flying over his crops can detect which are doing well and which need additional care. This is done by a sensor that can detect the difference in color and analyze its meaning. Each plant has a different "signature," meaning that its color changes with the balance of chlorophylls, carotenoids, and water content.

Many hobbyist drones can now be fitted with sensors such as sonar or LIDAR that can detect the drone's height or an object. Although this technology has been around for a number of years, I have yet to see it work more than half the time. Yet it will come. Future drones will have sense-and-avoid capabilities built in. GeoFencing coordinates or using beacons around a specific place will prevent unauthorized drones from entering.

ANTI-DRONE

There has been an anti-drone program around for several years. That is to say, one drone or a device brings down an intruder drone. Although there is sense in this—especially when we are talking about airports or government installations—the realities have yet to be realized. Various methods have been tried that range from attack drones which carry a net and simply scoop the offending drone, to high-powered microwaves that will take out the drone's electronics—and most likely half a dozen innocent people as well.

There are now drones aimed at protecting nuclear power plants by dropping something into the rotors of any intruder. In Japan, an anti-nuclear protester landed a drone bearing a radiation hazard symbol on the roof of the Prime Minister's office that gave off a non-harmful trace of radiation. In a similar vein, a drone crashed onto the White House lawn in 2015, showing that no one is safe. One method of detecting a drone is to listen for its noise signature, which in any drone is pretty loud. This requires sensors being placed on buildings that will listen, analyze, and warn. Although this is a step in the right direction, you still have to visually identify the drone and bring it down—safely. There is no point in knocking it out of the sky only for it to fall on innocent bystanders.

No matter how we approach the subject, defeating drones is one problem that many governments are keen on solving, the main reason being terrorism. For a long time, the weaponizing of drones, other than the larger military type, has been put on the back burner. However, in the past year, several countries, including the US, China, and India, have seriously looked at using small drones for military use. This may seem close to a *Star Wars* mentality, but the truth is that it can be done—and therefore it will be done. Why put soldiers in harm's way when we can put a drone there instead is one logical argument, but the devastation of going down this road will still have a severe impact. Imagine troops moving over open ground only to be confronted by swarms of small aerial mine-carrying drones. Just another reason why there needs to be a serious anti-drone program.

SUMMARY

It is only 112 years (December 7, 1903) since the Wright Brothers' first flight when Orville piloted the first powered airplane 20 feet above the ground. The flight lasted just 12 seconds and flew a total of 120 feet, and later that day they managed 59 seconds with a distance of 852 feet. Today, a $20 model toy drone can do just the same. This rate of advancement—coupled with newer technologies—will push drones way past what we think they can achieve. Innovators are already making drones that will support a person, so how long before we are flying the skies sitting in our own drone? It is the marvel of our age, and it fills me personally with hope and a wish to come back in 112 years from now and see how far we have progressed.

GLOSSARY

This is a glossary from FlyingTech UK of some of the drone-related terms you may hear or need to know or understand.

AHRS: Attitude and Heading Reference System.

APM: ArduPilotMega autopilot electronics.

ArduCopter: Rotary-wing autopilot software for the APM and Pixhawk electronics.

ArduPlane: Fixed-wing autopilot software for the APM and Pixhawk electronics.

ArduPilot: The overall autopilot project that ArduCopter, ArduPlane, and ArduRover live within.

ArduRover: Ground and water autopilot software for the APM and Pixhawk electronics.

Arduino: An open source embedded processor project. Includes a hardware standard originally based on the Atmel Atmega (and other 8-bit) microprocessor microcontroller and necessary supporting hardware, and a software programming environment based on the C-like Processing language.

BEC: Battery Elimination Circuit. A voltage regulator found in ESCs (see entry below) and as a stand-alone product. Designed to provide constant 5v voltage for RC equipment, autopilots, and other onboard electronics.

BASIC Stamp: A simple embedded processor controller and programming environment created and sold by Parallax. Often used to teach basic embedded computing and the basis of our autopilot tutorial project. Parallax also makes the very capable Propeller chip.

Bluetooth: A wireless technology standard for exchanging data over short distances (using UHF radio waves in the ISM band from 2.4 to 2.485 GHz) from fixed and mobile devices, and building Personal Area Networks (PANs). Originally conceived as a wireless alternative to RS-232 data cables. It can connect several concurrent devices.

Bootloader: Special code stored in non-volatile memory in a microprocessor that can interface with a PC to download a user's program.

COA: Certificate of Authorization. FAA approval for a UAV flight.

Eagle file: The schematic and PCB design files (and related files that tell PCB fabricators how to create the boards) generated by the free Cadsoft Eagle program. This is the most common standard used in the open source hardware world, although, ironically, it's not open source software itself. Needless to say, this is not optimal, and the Eagle software is clumsy and hard to learn. One hopes that an open source alternative will someday emerge.

DCM: Direction Cosine Matrix. An algorithm that is a less processing-intensive equivalent of the Kalman Filter.

DSM/DSM2/DSMX: Spektrum, an RC equipment maker, refers to its proprietary technology as "Digital Spectrum Modulation." Each transmitter has a globally unique identifier (GUID), to which receivers can be bound, ensuring that no transmitter will interfere with other nearby Spektrum DSM systems. DSM uses Direct-Sequence Spread Spectrum (DSSS) technology.

DSSS: Direct-Sequence Spread Spectrum is a modulation technique. As with other spread spectrum technologies, the transmitted signal takes up more bandwidth than the information signal that modulates the carrier or broadcast frequency. The name "spread spectrum" comes from the fact that the carrier signals occur over the full bandwidth (spectrum) of a device's transmitting frequency.

EEPROM: Electronically Erasable Programmable Read Only Memory. A type of non-volatile memory used in computers and other electronic devices to store small amounts of data that must be saved when power is removed, for example, static calibration/reference tables. Unlike bytes in most other kinds of non-volatile memory, individual bytes in a traditional EEPROM can be independently read, erased, and rewritten.

ESC: Electronic Speed Control. Device to control the motor in an electric aircraft. Serves as the connection between the main battery and the RC receiver. Usually includes a Battery Elimination Circuit (BEC), which provides power for the RC system and other onboard electronics, such as an autopilot.

FREQUENCIES: The most common frequencies used for video transmission are 900 MHz, 1.2 GHz, 2.4 GHz, and 5.8 GHz. Specialized long-range UHF control systems operating at 433 MHz or 869 MHz are commonly used to achieve greater control range, while the use of directional, high-gain antennas increases video range. Sophisticated setups are capable of achieving a range of 20–30 miles or more.

FHSS Frequency-Hopping Spread Spectrum is a method of transmitting radio signals by rapidly switching a carrier among many frequency channels, using a pseudorandom sequence known to both transmitter and receiver. Advantages over a fixed-frequency transmission: 1. Spread-spectrum signals are highly

resistant to narrowband interference. The process of re-collecting a spread signal spreads out the interfering signal, causing it to recede into the background.
2. Spread-spectrum signals are difficult to intercept. A spread-spectrum signal
may simply appear as an increase in the background noise to a narrowband
receiver. An eavesdropper may have difficulty intercepting a transmission in real
time if the pseudorandom sequence is not known. 3. Spread-spectrum transmissions can share a frequency band with many types of conventional transmissions
with minimal interference. The spread-spectrum signals add minimal noise to
the narrow-frequency communications, and vice versa. As a result, bandwidth
can be used more efficiently.

FPV: First person view. A technique that uses an onboard video camera and wireless connection to the ground that allows a pilot on the ground with video goggles
to fly with a cockpit view.

FTDI: Future Technology Devices International, which is the name of the company that makes the chips. A standard to convert USB to serial communications.
Available as a chip for boards that have a USB connector, or in a cable to connect
to breakout pins.

GCS: Ground Control Station. Software running on a computer on the ground that
receives telemetry information from an airborne UAV and displays its progress
and status, often including video and other sensor data. Can also be used to
transmit in-flight commands to the UAV.

GIT: A version control system for software developers. The DIY Drones team uses
a Git-based service called GitHub.

Hardware-in-the-loop simulation: Doing a simulation where software running
on another computer generates data that simulates the data that would be coming from an autopilot's sensors. The autopilot is running and doesn't "know" that
the data is simulated, so it responds just as it would to real sensor data. Hardware-
in-the-loop uses the physical autopilot hardware connected to a simulator, as
opposed to simulating the autopilot in software, too.

I2C: Inter-Integrated Circuit. A serial bus that allows multiple low speed peripherals, such as sensors, to be connected to a microprocessor.

IDE: An Integrated Development Environment, such as the Arduino editor/down-
loader/serial monitor software. Often includes a debugger.

IMU: Inertial Measurement Unit. Usually has at least three accelerometers
(measuring the gravity vector in the X, Y, and Z dimensions) and two gyros (measuring rotation around the tilt and pitch axis). Neither is sufficient by itself, since
accelerometers are thrown off by movement (that is, they are "noisy" over short

periods of time), whereas gyros drift over time. The data from both types of sensors must be combined in software to determine true aircraft attitude and movement. One technique for doing this is the Kalman filter (see entry).

Inner loop/Outer loop: Usually used to refer to the stabilization and navigation functions of an autopilot. The stabilization function must run in real time and as often as 100 times a second ("inner loop"), while the navigation function can run as infrequently as once per second and can tolerate delays and interruptions ("outer loop").

INS: Inertial Navigation System. A way to calculate position based on an initial GPS reading followed by readings from motion and speed sensors. Useful when GPS is not available or has temporarily lost its signal.

ICSP: In Circuit Serial Programmer. A way to load code to a microprocessor microcontroller. Usually seen as a six-pin (two rows of three) connector on a PCB. To use this, you need a programmer, such as this one, that uses the SPI (Serial Peripheral Interface) standard.

Kalman filter: A relatively complicated algorithm that is primarily used to combine accelerometer and gyro data to provide an accurate description of aircraft attitude and movement in real time.

LIDAR: Light Detection and Ranging. A remote sensing technology that collects 3-dimensional point clouds of the Earth's surface. The unit attached to the front of the SQ-4 can detect objects as small as 1 cm up to a range of 10 m. Its low power consumption and weight make it ideal for small quad-copters.

With the implementation of a new signal processing architecture, the SQ-4 LIDAR will now operate at measurement speeds of up to 500 readings per second, offering greater resolution for scanning applications.

Whereas traditional LIDAR has been used to measure distance from the ground, the SQ-4 uses the same technique to sense and avoid objects. It does this by integrating directly with the BCB autopilot and not the traditional method of connecting to the I2C Communications.

The LIDAR on the SQ-4 is switched on and off automatically whenever the drone is flown in manual mode and the function AltHold is activated. Switching to any other flight mode such as Loiter, Stabilized, or Automatic will deactivate the LIDAR.

When LIDAR is active and it senses an object in its forward path, the drone will stop. If the object advances towards the SQ-4 (moving human or vehicle), then the SQ-4 will retreat automatically.

LOS: Line of Sight. Refers to an FAA requirement that UAVs stay within pilots' direct visual control if they are flying under the recreational exemption to COA approval.

LiPo: Lithium polymer battery, also called LiPoly. Variants include Lithium ion (Li-Ion) battery. This battery chemistry offers more power and lighter weight than NiMh and NiCad batteries.

MAV: Micro Air Vehicle. A small UAV.

MAVLink: The Micro Air Vehicle communications Link protocol used by the ArduCopter and ArduPlane line of autopilots.

Microprocessor: A microprocessor incorporates the functions of a computer's central processing unit (CPU) on a single integrated circuit or at most, a few integrated circuits (system clock, memory, peripheral device drivers).

Microcontroller: A microcontroller (sometimes abbreviated µC, uC or MCU) is a small computer on a single integrated circuit containing a processor core, memory, and programmable input/output peripherals. Program memory in the form of flash or EEPROM is included on the chip, as well as a typically small amount of RAM. Microcontrollers are designed for embedded applications, in contrast to the microprocessors used in personal computers or other general purpose applications.

NMEA: National Marine Electronics Association standard for GPS information. When we refer to "NMEA sentences," we're talking about ASCII strings from a GPS module that look like this: $GPGGA,123519,4807.038,N,01131.000,E,1,08,0.9 ,545.4,M,46.9,M,*47

OSD: On-Screen Display. A way to integrate data (often telemetry information) into the real-time video stream the aircraft is sending to the ground.

PCB: Printed Circuit Board. In our use, a specialized board designed and "fabricated" for a dedicated purpose, as opposed to a breadboard or prototype board, which can be used and reused for many projects.

PCM: Pulse Coded Modulation. A method used to digitally represent sampled analog signals. It is the standard form of digital audio in computers, CDs, digital telephony, and other digital audio applications. In a PCM stream, the amplitude of the analog signal is sampled regularly at uniform intervals, and each sample is quantized to the nearest value within a range of digital steps. Primarily useful for optical communications systems, where there tends to be little or no multipath interference.

PIC: Pilot in Command. Refers to an FAA requirement that UAVs stay under a pilot's direct control if they are flying under the recreational exemption to COA approval. See the entry LOS for more.

PID: Proportional/Integral/Derivative control method. A machine control algorithm that allows for more accurate sensor-motion control loops and less over-control.

Pixhawk: The next-gen 32-bit autopilot, which succeeded APM. A collaboration between 3D Robotics and the PX4 team at ETH, the technical university in Zurich.

POI: Point of Interest, also known as Region of Interest. Designates a spot that a UAV should keep a camera pointed towards.

PPM: Pulse Position Modulation. Signal modulation in which a set number of message bits are encoded by transmitting a single pulse in one of possible 2(number if message bits) time-shifts.

PWM: Pulse Width Modulation. The square-wave signals used in RC control to drive servos and speed controllers.

ROI: Region of Interest. Also known as Point of Interest (see the entry POI).

RTL: Return to Launch. Return the aircraft to the "home" position where it took off.

Shield: a specialized board that fits on top of an Arduino to add a specific function, such as wireless data or GPS.

SiRF III: SiRF is a technology company that has developed a standard used by most modern GPS modules. Includes SiRF III binary mode, which is an alternative to the ASCII-based NMEA standard (see entry).

Sketch: The program files, drivers, and other code generated by the Arduinio IDE for a single project.

SVN: Subversion Version-control Number repository, used by the DIY Drones (in the past) and other teams for source code.

Thermopile: An infrared detector. Often used in pairs in UAVs to measure tilt and pitch by looking at differences in the infrared signature of the horizon fore and aft and on both sides. This is based on the fact that there is always an infrared gradient between earth and sky, and that you can keep a plane flying level by ensuring that the readings are the same from both sensors in each pair, each looking in opposite directions.

UAV: Unmanned Aerial Vehicle. In the military, these are increasingly called Unmanned Aerial Systems (UAS), to reflect that the aircraft is just part of a complex system in the air and on the ground. Ground-based autonomous robots are called Unmanned Ground Vehicles (UGVs) and robot submersibles are called Autonomous Underwater Vehicles (AUVs). Robot boats are called Unmanned Surface Vehicles (USVs).

WAAS: Wide Area Augmentation System. A system of satellites and ground stations that provide GPS signal corrections, giving up to five times better position accuracy than uncorrected GPS.

ZigBee: A wireless communications standard, which has longer range than Bluetooth but lower power consumption than Wi-Fi.

Author's Note: It is possible to take off in AltHold with LIDAR activated, and this mode is recommended when flying in a restricted area such as a forest. When working in obstacle avoidance, it is best to restrict the SQ-4's speed taking into account the direction of travel inertia movement.

ACKNOWLEDGMENTS

This book would not have been possible had it not been for all the help, advice, and permissions I have received from so many people. The list is long but they include Nuno Simoes of UAVision, Chris Anderson of 3DR, Steven Prior from Southampton University, my friend Tim Whitcombe, who taught me how to fly safely, and Ivan Reedman of Torquing Robotics. The help from so many when I bombarded them through DIY Drones and other drone websites was immense. Finally, thanks to the robotics engineer Rhys Isaac at BCB, who checked over the important construction chapters, William at FlyingTech for permission and help with the Glossary, and Marija Ullman for the wonderfully rapid proofreading.

INDEX

Bold page numbers indicate photograph on that page.

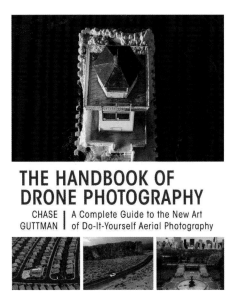

The Handbook of Drone Photography
A Complete Guide to the New Art of Do-It-Yourself Aerial Photography

by Chase Guttman

A Beautifully Photographed Guide on How to Use Drone Technology to Capture Aerial Photographs and Videos.

Drones are the next frontier in photography and videography. This cutting-edge technology, still unexplored by the masses, can bring visual artistry to new and exciting heights. *The Handbook of Drone Photography* will be the go-to manual for consumers wishing to harness the power of drones to capture stunning aerial photographs and gorgeous high-altitude videos. From sweeping vistas of distant lands to fun stills of family and friends in the backyard, drones can capture it all. With this powerful tool, photographs and videos can transcend the ordinary and ascend to the extraordinary. This book will put the world of drones and the future of photography and videography in readers' hands.

The Handbook of Drone Photography will cover everything one needs to choose the right drone, get airborne and capture and share incredible content. With easy and straightforward instruction, the book will familiarize readers with their craft and its controls. Readers will learn to master drones' extraordinary image capturing capabilities, review detailed photography and videography tips that can bring their artistic vision to life and help them discover post-processing techniques that will make the quality of their work soar. For the first time, aerial photography is open to everyone and award-winning travel photographer Chase Guttman will guide readers' drone ventures from beginning to end. The wide array of possibilities from this untapped medium are just now being explored. Readers will get in on the ground floor so that they can be seen as innovators as this technology rapidly popularizes. *The Handbook of Drone Photography* can help anyone break into this thrilling, high-potential space and launch their own lofty explorations today.

$17.99 Paperback ISBN 978-1-5107-1216-4

ENJOY OTHER BOOKS BY BARRY DAVIES!

SAS Combat Handbook
ISBN: 978-1-63220-295-6
Price: $14.99

SAS Tracking Handbook
ISBN: 978-1-62914-235-7
Price: $14.95

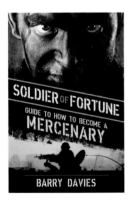

Soldier of Fortune Guide
to How to Become a
Mercenary
ISBN: 978-1-62087-097-6
Price: $14.95

Soldier of Fortune Guide
to How to Disappear and
Never Be Found
ISBN: 978-1-62087-787-6
Price: $14.95